数据科学原理(影印版)
Principles of Data Science

Sinan Ozdemir 著

南京 东南大学出版社

图书在版编目(CIP)数据

数据科学原理:英文/(美)思南·约茨德米尔(Sinan Ozdemir)著.—影印本.—南京:东南大学出版社,2017.10(2018.10重印)

书名原文:Principles of Data Science

ISBN 978-7-5641-7364-7

Ⅰ.①数… Ⅱ.①思… Ⅲ.①数据处理-英文 Ⅳ.①TP274

中国版本图书馆CIP数据核字(2017)第196705号

图字:10-2017-119号

数据科学原理(影印版)

出版发行:东南大学出版社
地　　址:南京四牌楼2号　　邮编:210096
出 版 人:江建中
网　　址:http://www.seupress.com
电子邮件:press@seupress.com
印　　刷:江苏凤凰数码印务有限公司
开　　本:787毫米×980毫米　16开本
印　　张:24.25
字　　数:475千字
版　　次:2017年10月第1版
印　　次:2018年10月第2次印刷
书　　号:ISBN 978-7-5641-7364-7
定　　价:92.00元

本社图书若有印装质量问题,请直接与营销部联系。电话(传真):025-83791830

Credits

Author
Sinan Ozdemir

Reviewers
Samir Madhavan
Oleg Okun

Acquisition Editor
Sonali Vernekar

Content Development Editor
Samantha Gonsalves

Technical Editor
Anushree Arun Tendulkar

Copy Editor
Shaila Kusanale

Project Coordinator
Devanshi Doshi

Proofreaders
Safis Editing

Indexer
Tejal Daruwale Soni

Graphics
Jason Monteiro

Production Coordinator
Melwyn Dsa

Cover Work
Melwyn Dsa

About the Author

Sinan Ozdemir is a data scientist, startup founder, and educator living in the San Francisco Bay Area with his dog, Charlie; cat, Euclid; and bearded dragon, Fiero. He spent his academic career studying pure mathematics at Johns Hopkins University before transitioning to education. He spent several years conducting lectures on data science at Johns Hopkins University and at the General Assembly before founding his own start-up, Legion Analytics, which uses artificial intelligence and data science to power enterprise sales teams.

After completing the Fellowship at the Y Combinator accelerator, Sinan has spent most of his days working on his fast-growing company, while creating educational material for data science.

I would like to thank my parents and my sister for supporting me through life, and also, my various mentors, including Dr. Pam Sheff of Johns Hopkins University and Nathan Neal, the chapter adviser of my collegiate leadership fraternity, Sigma Chi.

Thank you to Packt Publishing for giving me this opportunity to share the principles of data science and my excitement for how this field will impact all of our lives in the coming years.

About the Reviewers

Samir Madhavan has over six years of rich data science experience in the industry and has also written a book called *Mastering Python for Data Science*. He started his career with Mindtree, where he was a part of the fraud detection algorithm team for the UID (Unique Identification) project, called Aadhar, which is the equivalent of a Social Security number for India. After this, he joined Flutura Decision Sciences and Analytics as the first employee, where he was part of the core team that helped the organization scale to an over a hundred members. As a part of Flutura, he helped establish big data and machine learning practice within Flutura and also helped out in business development. At present, he is leading the analytics team for a Boston-based pharma tech company called Zapprx, and is helping the firm to create data-driven products that will be sold to its customers.

Oleg Okun is a machine learning expert and an author/editor of four books, numerous journal articles, and conference papers. His career spans more than a quarter of a century. He was employed in both academia and industry in his mother country, Belarus, and abroad (Finland, Sweden, and Germany). His work experience includes document image analysis, fingerprint biometrics, bioinformatics, online/offline marketing analytics, and credit-scoring analytics.

He is interested in all aspects of distributed machine learning and the Internet of Things. Oleg currently lives and works in Hamburg, Germany.

I would like to express my deepest gratitude to my parents for everything that they have done for me.

www.PacktPub.com

eBooks, discount offers, and more

Did you know that Packt offers eBook versions of every book published, with PDF and ePub files available? You can upgrade to the eBook version at www.PacktPub.com and as a print book customer, you are entitled to a discount on the eBook copy. Get in touch with us at customercare@packtpub.com for more details.

At www.PacktPub.com, you can also read a collection of free technical articles, sign up for a range of free newsletters and receive exclusive discounts and offers on Packt books and eBooks.

https://www.packtpub.com/mapt

Get the most in-demand software skills with Mapt. Mapt gives you full access to all Packt books and video courses, as well as industry-leading tools to help you plan your personal development and advance your career.

Why subscribe?

- Fully searchable across every book published by Packt
- Copy and paste, print, and bookmark content
- On demand and accessible via a web browser

Table of Contents

Preface vii

Chapter 1: How to Sound Like a Data Scientist 1

What is data science? 3
Basic terminology 3
Why data science? 5
Example – Sigma Technologies 5
The data science Venn diagram 6
The math 8
Example – spawner-recruit models 8
Computer programming 10
Why Python? 10
Python practices 11
Example of basic Python 12
Domain knowledge 14
Some more terminology 15
Data science case studies 16
Case study – automating government paper pushing 16
Fire all humans, right? 18
Case study – marketing dollars 18
Case study – what's in a job description? 20
Summary 23

Chapter 2: Types of Data 25

Flavors of data 25
Why look at these distinctions? 26
Structured versus unstructured data 26
Example of data preprocessing 27
Word/phrase counts 28
Presence of certain special characters 28
Relative length of text 29
Picking out topics 29

Quantitative versus qualitative data 30
Example – coffee shop data 30
Example – world alcohol consumption data 32
Digging deeper 34
The road thus far... 34
The four levels of data 35
The nominal level 35
Mathematical operations allowed 36
Measures of center 36
What data is like at the nominal level 36
The ordinal level 36
Examples 37
Mathematical operations allowed 37
Measures of center 38
Quick recap and check 39
The interval level 39
Example 39
Mathematical operations allowed 40
Measures of center 40
Measures of variation 41
The ratio level 43
Examples 43
Measures of center 43
Problems with the ratio level 44
Data is in the eye of the beholder 45
Summary 45
Chapter 3: The Five Steps of Data Science 47
Introduction to Data Science 47
Overview of the five steps 48
Ask an interesting question 48
Obtain the data 48
Explore the data 48
Model the data 49
Communicate and visualize the results 49
Explore the data 49
Basic questions for data exploration 50
Dataset 1 – Yelp 51
Dataframes 53
Series 54
Exploration tips for qualitative data 54
Dataset 2 – titanic 60
Summary 64

Chapter 4: Basic Mathematics **65**
Mathematics as a discipline **65**
Basic symbols and terminology **66**
Vectors and matrices 66
Quick exercises 68
Answers 69
Arithmetic symbols 69
Summation 69
Proportional 70
Dot product 70
Graphs 73
Logarithms/exponents 74
Set theory 77
Linear algebra **81**
Matrix multiplication 81
How to multiply matrices 82
Summary **85**
Chapter 5: Impossible or Improbable – A Gentle Introduction to Probability **87**
Basic definitions **88**
Probability **88**
Bayesian versus Frequentist **90**
Frequentist approach 90
The law of large numbers 91
Compound events **93**
Conditional probability **96**
The rules of probability **97**
The addition rule 97
Mutual exclusivity 98
The multiplication rule 99
Independence 100
Complementary events 100
A bit deeper **102**
Summary **103**
Chapter 6: Advanced Probability **105**
Collectively exhaustive events **105**
Bayesian ideas revisited **106**
Bayes theorem 106
More applications of Bayes theorem 110
Example – Titanic 110
Example – medical studies 112

Random variables 113
Discrete random variables 114
Types of discrete random variables 119
Summary 128
Chapter 7: Basic Statistics 131
What are statistics? 131
How do we obtain and sample data? 133
Obtaining data 133
Observational 133
Experimental 133
Sampling data 136
Probability sampling 136
Random sampling 136
Unequal probability sampling 137
How do we measure statistics? 138
Measures of center 138
Measures of variation 139
Definition 144
Example – employee salaries 144
Measures of relative standing 145
The insightful part – correlations in data 151
The Empirical rule 153
Summary 155
Chapter 8: Advanced Statistics 157
Point estimates 157
Sampling distributions 162
Confidence intervals 164
Hypothesis tests 168
Conducting a hypothesis test 169
One sample t-tests 170
Example of a one sample t-tests 170
Assumptions of the one sample t-tests 171
Type I and type II errors 174
Hypothesis test for categorical variables 174
Chi-square goodness of fit test 175
Chi-square test for association/independence 177
Summary 180
Chapter 9: Communicating Data 181
Why does communication matter? 181
Identifying effective and ineffective visualizations 182
Scatter plots 182
Line graphs 184

Bar charts 185
Histograms 187
Box plots 189
When graphs and statistics lie 191
Correlation versus causation 192
Simpson's paradox 195
If correlation doesn't imply causation, then what does? 196
Verbal communication 196
It's about telling a story 197
On the more formal side of things 197
The why/how/what strategy of presenting 198
Summary 199
Chapter 10: How to Tell If Your Toaster Is Learning – Machine Learning Essentials 201
What is machine learning? 202
Machine learning isn't perfect 204
How does machine learning work? 205
Types of machine learning 205
Supervised learning 206
It's not only about predictions 209
Types of supervised learning 209
Data is in the eyes of the beholder 211
Unsupervised learning 212
Reinforcement learning 214
Overview of the types of machine learning 215
How does statistical modeling fit into all of this? 217
Linear regression 217
Adding more predictors 222
Regression metrics 224
Logistic regression 231
Probability, odds, and log odds 233
The math of logistic regression 236
Dummy variables 239
Summary 244
Chapter 11: Predictions Don't Grow on Trees – or Do They? 245
Naïve Bayes classification 245
Decision trees 254
How does a computer build a regression tree? 256
How does a computer fit a classification tree? 256
Unsupervised learning 262
When to use unsupervised learning 262

K-means clustering **262**
Illustrative example – data points 264
Illustrative example – beer! 270
Choosing an optimal number for K and cluster validation **273**
The Silhouette Coefficient 273
Feature extraction and principal component analysis **276**
Summary **287**
Chapter 12: Beyond the Essentials **289**
The bias variance tradeoff **290**
Error due to bias 290
Error due to variance 290
Two extreme cases of bias/variance tradeoff 298
Underfitting 298
Overfitting 299
How bias/variance play into error functions 299
K folds cross-validation **301**
Grid searching **305**
Visualizing training error versus cross-validation error 308
Ensembling techniques **310**
Random forests 312
Comparing Random forests with decision trees 317
Neural networks **318**
Basic structure 318
Summary **324**
Chapter 13: Case Studies **325**
Case study 1 – predicting stock prices based on social media **325**
Text sentiment analysis 325
Exploratory data analysis 326
Regression route 337
Classification route 340
Going beyond with this example 342
Case study 2 – why do some people cheat on their spouses? **342**
Case study 3 – using tensorflow **350**
Tensorflow and neural networks 354
Summary **361**
Index **363**

Preface

The topic of this book is data science, which is a field of study and application that has been growing rapidly for the past several decades. As a growing field, it is gaining a lot of attention in both the media as well as in the job market. The United States recently appointed its first ever chief data scientist, DJ Patil. This move was modeled after tech companies who, honestly, only recently started hiring massive data teams. These skills are in high demand and their applications extend much further than today's job market.

This book will attempt to bridge the gap between math/programming/domain expertise. Most people today have expertise in at least one of these (maybe two), but proper data science requires a little bit of all three. We will dive into topics from all three areas and solve complex problems. We will clean, explore, and analyze data in order to derive scientific and accurate conclusions. Machine learning and deep learning techniques will be applied to solve complex data tasks.

What this book covers

Chapter 1, How to Sound Like a Data Scientist, gives an introduction to the basic terminology used by data scientists and a look at the types of problem we will be solving throughout this book.

Chapter 2, Types of Data, looks at the different levels and types of data out there and how to manipulate each type. This chapter will begin to deal with the mathematics needed for data science.

Chapter 3, The Five Steps of Data Science, uncovers the five basic steps of performing data science, including data manipulation and cleaning, and sees examples of each step in detail.

Chapter 4, Basic Mathematics, helps us discover the basic mathematical principles that guide the actions of data scientists by seeing and solving examples in calculus, linear algebra, and more.

Chapter 5, Impossible or Improbable – a Gentle Introduction to Probability, is a beginner's look into probability theory and how it is used to gain an understanding of our random universe.

Chapter 6, Advanced Probability, uses principles from the previous chapter and introduces and applies theorems, such as the Bayes Theorem, in the hope of uncovering the hidden meaning in our world.

Chapter 7, Basic Statistics, deals with the types of problem that statistical inference attempts to explain, using the basics of experimentation, normalization, and random sampling.

Chapter 8, Advanced Statistics, uses hypothesis testing and confidence interval in order to gain insight from our experiments. Being able to pick which test is appropriate and how to interpret p-values and other results is very important as well.

Chapter 9, Communicating Data, explains how correlation and causation affect our interpretation of data. We will also be using visualizations in order to share our results with the world.

Chapter 10, How to Tell If Your Toaster Is Learning – Machine Learning Essentials, focuses on the definition of machine learning and looks at real-life examples of how and when machine learning is applied. A basic understanding of the relevance of model evaluation is introduced.

Chapter 11, Predictions Don't Grow on Trees, or Do They?, looks at more complicated machine learning models, such as decision trees and Bayesian-based predictions, in order to solve more complex data-related tasks.

Chapter 12, Beyond the Essentials, introduces some of the mysterious forces guiding data sciences, including bias and variance. Neural networks are introduced as a modern deep learning technique.

Chapter 13, Case Studies, uses an array of case studies in order to solidify the ideas of data science. We will be following the entire data science workflow from start to finish multiple times for different examples, including stock price prediction and handwriting detection.

What you need for this book

This book uses Python to complete all of its code examples. A machine (Linux/Mac/Windows OK) with access to a Unix-style Terminal with Python 2.7 installed is required. Installation of the Anaconda distribution is also recommended as it comes with most of the packages used in the examples.

Who this book is for

This book is for people who are looking to understand and utilize the basic practices of data science for any domain.

The reader should be fairly well acquainted with basic mathematics (algebra, perhaps probabilities) and should feel comfortable reading snippets of R/Python as well as pseudocode. The reader is not expected to have worked in a data field; however, they should have the urge to learn and apply the techniques put forth in this book to either their own datasets or those provided to them.

Conventions

In this book, you will find a number of text styles that distinguish between different kinds of information. Here are some examples of these styles and an explanation of their meaning.

Code words in text, database table names, folder names, filenames, file extensions, pathnames, dummy URLs, user input, and Twitter handles are shown as follows: "For these operators, keep the `boolean` datatype in mind."

A block of code is set as follows:

```
tweet = "RT @j_o_n_dnger: $TWTR now top holding for
             Andor, unseating $AAPL"

words_in_tweet = first_tweet.split(' ') # list of words in tweet
```

When we wish to draw your attention to a particular part of a code block, the relevant lines or items are set in bold:

```
for word in words_in_tweet:              # for each word in list
  if "$" in word:                        # if word has a "cashtag"
  print "THIS TWEET IS ABOUT", word  # alert the user
```

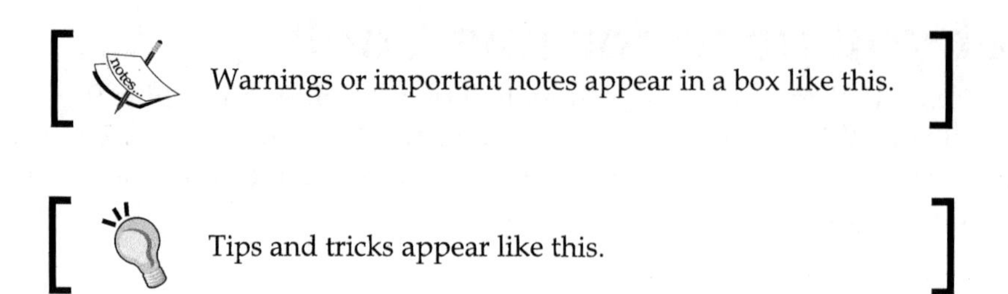

Warnings or important notes appear in a box like this.

Tips and tricks appear like this.

Reader feedback

Feedback from our readers is always welcome. Let us know what you think about this book—what you liked or disliked. Reader feedback is important for us as it helps us develop titles that you will really get the most out of.

To send us general feedback, simply e-mail `feedback@packtpub.com`, and mention the book's title in the subject of your message.

If there is a topic that you have expertise in and you are interested in either writing or contributing to a book, see our author guide at `www.packtpub.com/authors`.

Customer support

Now that you are the proud owner of a Packt book, we have a number of things to help you to get the most from your purchase.

Downloading the example code

You can download the example code files for this book from your account at `http://www.packtpub.com`. If you purchased this book elsewhere, you can visit `http://www.packtpub.com/support` and register to have the files e-mailed directly to you.

You can download the code files by following these steps:

1. Log in or register to our website using your e-mail address and password.
2. Hover the mouse pointer on the **SUPPORT** tab at the top.
3. Click on **Code Downloads & Errata**.

4. Enter the name of the book in the **Search** box.
5. Select the book for which you're looking to download the code files.
6. Choose from the drop-down menu where you purchased this book from.
7. Click on **Code Download**.

You can also download the code files by clicking on the **Code Files** button on the book's webpage at the Packt Publishing website. This page can be accessed by entering the book's name in the **Search** box. Please note that you need to be logged in to your Packt account.

Once the file is downloaded, please make sure that you unzip or extract the folder using the latest version of:

- WinRAR / 7-Zip for Windows
- Zipeg / iZip / UnRarX for Mac
- 7-Zip / PeaZip for Linux

The code bundle for the book is also hosted on GitHub at `https://github.com/PacktPublishing/Principles-of-Data-Science`. We also have other code bundles from our rich catalog of books and videos available at `https://github.com/PacktPublishing/`. Check them out!

Downloading the color images of this book

We also provide you with a PDF file that has color images of the screenshots/diagrams used in this book. The color images will help you better understand the changes in the output. You can download this file from `https://www.packtpub.com/sites/default/files/downloads/PrinciplesofDataScience_ColorImages.pdf`.

Errata

Although we have taken every care to ensure the accuracy of our content, mistakes do happen. If you find a mistake in one of our books—maybe a mistake in the text or the code—we would be grateful if you could report this to us. By doing so, you can save other readers from frustration and help us improve subsequent versions of this book. If you find any errata, please report them by visiting `http://www.packtpub.com/submit-errata`, selecting your book, clicking on the **Errata Submission Form** link, and entering the details of your errata. Once your errata are verified, your submission will be accepted and the errata will be uploaded to our website or added to any list of the existing errata under the Errata section of that title.

To view the previously submitted errata, go to `https://www.packtpub.com/books/content/support` and enter the name of the book in the search field. The required information will appear under the **Errata** section.

Piracy

Piracy of copyrighted material on the Internet is an ongoing problem across all media. At Packt, we take the protection of our copyright and licenses very seriously. If you come across any illegal copies of our works in any form on the Internet, please provide us with the location address or website name immediately so that we can pursue a remedy.

Please contact us at `copyright@packtpub.com` with a link to the suspected pirated material.

We appreciate your help in protecting our authors and our ability to bring you valuable content.

Questions

If you have a problem with any aspect of this book, you can contact us at `questions@packtpub.com`, and we will do our best to address the problem.

1
How to Sound Like a Data Scientist

No matter which industry you work in, IT, fashion, food, or finance, there is no doubt that data affects your life and work. At some point in this week, you will either have or hear a conversation about data. News outlets are covering more and more stories about data leaks, cybercrimes, and how data can give us a glimpse into our lives. But why now? What makes this era such a hotbed for data-related industries?

In the 19th century, the world was in the grip of the *industrial age*. Mankind was exploring its place in industry alongside giant mechanical inventions. Captains of industry, such as Henry Ford, recognized major market opportunities at the hands of these machines, and were able to achieve previously unimaginable profits. Of course the industrial age had its pros and cons. While mass production placed goods in the hands of more consumers, our battle with pollution also began around this time.

By the 20th century, we were quite skilled at making huge machines; the goal now was to make them smaller and faster. The industrial age was over and was replaced by what we refer to as the *information age*. We started using machines to gather and store information (data) about ourselves and our environment for the purpose of understanding our universe.

Beginning in the 1940s, machines like **ENIAC** (considered one of, if not the first, computer) were computing math equations and running models and simulations like never before.

The ENIAC, http://ftp.arl.mil/ftp/historic-computers/

We finally had a decent lab assistant who could run the numbers better than we could! As with the industrial age, the information age brought us both the good and the bad. The good was the extraordinary pieces of technology, including mobile phones and televisions. The bad in this case was not as bad as worldwide pollution, but still left us with a problem in the 21st century, so much data.

That's right, the information age, in its quest to procure data, has exploded the production of electronic data. Estimates show that we created about 1.8 trillion gigabytes of data in 2011 (take a moment to just think about how much that is). Just one year later, in 2012, we created over 2.8 trillion gigabytes of data! This number is only going to explode further to hit an estimated 40 trillion gigabytes of data creation in just one year by 2020. People contribute to this every time they tweet, post on Facebook, save a new resume on Microsoft Word, or just send their mom a picture through text message.

Not only are we creating data at an unprecedented rate, we are consuming it at an accelerated pace as well. Just three years ago, in 2013, the average cell phone user used under 1 GB of data a month. Today, that number is estimated to be well over 2 GB a month. We aren't just looking for the next personality quiz, what we are looking for is insight. All of this data out there, some of it has to be useful to me! And it can be!

So we, in the 21st century, are left with a problem. We have so much data and we keep making more. We have built insanely tiny machines that collect data 24/7, and it's our job to make sense of it all. Enter the *data age*. This is the age when we take machines dreamed up by our 19th century ancestors and the data created by our 20th century counterparts and create insights and sources of knowledge that every human on Earth can benefit from. The United States created an entire new role in the government for the chief data scientist. Tech companies, such as Reddit, who up until now did not have a data scientist on their team, are now hiring them left and right. The benefit is quite obvious—using data to make accurate predictions and simulations gives us a look into our world like never before.

Sounds great, but what's the catch?

This chapter will explore the terminology and vocabulary of the modern data scientist. We will see key words and phrases that are essential in our discussion on data science throughout this book. We will also look at why we use data science and the three key domains data science is derived from before we begin to look at code in Python, the primary language used in this book:

- Basic terminology of data science
- The three domains of data science
- The basic Python syntax

What is data science?

Before we go any further, let's look at some basic definitions that we will use throughout this book. The great/awful thing about this field is that it is so young that these definitions can differ from textbook to newspaper to whitepaper.

Basic terminology

The definitions that follow are general enough to be used in daily conversations and work to serve the purpose of the book, an introduction to the principles of data science.

Let's start by defining what data is. This might seem like a silly first definition to have, but it is very important. Whenever we use the word "data", we refer to a collection of information in either an *organized* or *unorganized* format:

- **Organized data**: This refers to data that is sorted into a row/column structure, where every row represents a single *observation* and the columns represent the *characteristics* of that observation.
- **Unorganized data**: This is the type of data that is in the free form, usually text or raw audio/signals that must be parsed further to become organized.

  Whenever you open Excel (or any other spreadsheet program), you are looking at a blank row/column structure waiting for organized data. These programs don't do well with unorganized data. For the most part, we will deal with organized data as it is the easiest to glean insight from, but we will not shy away from looking at raw text and methods of processing unorganized forms of data.

Data science is the art and science of acquiring knowledge through data.

What a small definition for such a big topic, and rightfully so! Data science covers so many things that it would take pages to list it all out (I should know, I tried and got edited down).

Data science is all about how we take data, use it to acquire knowledge, and then use that knowledge to do the following:

- Make decisions
- Predict the future
- Understand the past/present
- Create new industries/products

This book is all about the methods of data science, including how to process data, gather insights, and use those insights to make informed decisions and predictions.

Data science is about using data in order to gain new insights that you would otherwise have missed.

As an example, imagine you are sitting around a table with three other people. The four of you have to make a decision based on some data. There are four opinions to consider. You would use data science to bring a fifth, sixth, and even seventh opinion to the table.

That's why data science won't replace the human brain, but complement it, work alongside it. Data science should not be thought of as an end-all solution to our data woes; it is merely an opinion, a very informed opinion, but an opinion nonetheless. It deserves a seat at the table.

Why data science?

In this data age, it's clear that we have a surplus of data. But why should that necessitate an entire new set of vocabulary? What was wrong with our previous forms of analysis? For one, the sheer volume of data makes it literally impossible for a human to parse it in a reasonable time. Data is collected in various forms and from different sources, and often comes in very unorganized.

Data can be missing, incomplete, or just flat out wrong. Often, we have data on very different scales and that makes it tough to compare it. Consider that we are looking at data in relation to pricing used cars. One characteristic of a car being the year it was made and another might be the number of miles on that car. Once we clean our data (which we spend a great deal of time looking at in this book), the relationships between the data become more obvious, and the knowledge that was once buried deep in millions of rows of data simply pops out. One of the main goals of data science is to make explicit practices and procedures to discover and apply these relationships in the data.

Earlier, we looked at data science in a more historical perspective, but let's take a minute to discuss its role in business today, through a very simple example.

Example – Sigma Technologies

Ben Runkle, CEO, Sigma Technologies, is trying to resolve a huge problem. The company is consistently losing long-time customers. He does not know why they are leaving, but he must do something fast. He is convinced that in order to reduce his churn, he must create new products and features, and consolidate existing technologies. To be safe, he calls in his chief data scientist, Dr. Jessie Hughan. However, she is not convinced that new products and features alone will save the company. Instead, she turns to the transcripts of recent customer service tickets. She shows Runkle the most recent transcripts and finds something surprising:

- ".... Not sure how to export this; are you?"
- "Where is the button that makes a new list?"
- "Wait, do you even know where the slider is?"
- "If I can't figure this out today, it's a real problem..."

It is clear that customers were having problems with the existing UI/UX, and weren't upset due to a lack of features. Runkle and Hughan organized a mass UI/UX overhaul and their sales have never been better.

Of course, the *science* used in the last example was minimal, but it makes a point. We tend to call people like Runkle, a driver. Today's common stick-to-your-gut CEO wants to make all decisions quickly and iterate over solutions until something works. Dr. Haghun is much more analytical. She wants to solve the problem just as much as Runkle, but she turns to user-generated data instead of her gut feeling for answers. Data science is about applying the skills of the analytical mind and using them as a driver would.

Both of these mentalities have their place in today's enterprises; however, it is Hagun's way of thinking that dominates the ideas of data science—using data generated by the company as her source of information rather than just picking up a solution and going with it.

The data science Venn diagram

It is a common misconception that only those with a PhD or geniuses can understand the math/programming behind data science. This is absolutely false. Understanding data science begins with three basic areas:

- **Math/statistics**: This is the use of equations and formulas to perform analysis
- **Computer programming**: This is the ability to use code to create outcomes on the computer
- **Domain knowledge**: This refers to understanding the problem domain (medicine, finance, social science, and so on)

The following Venn diagram provides a visual representation of how the three areas of data science intersect:

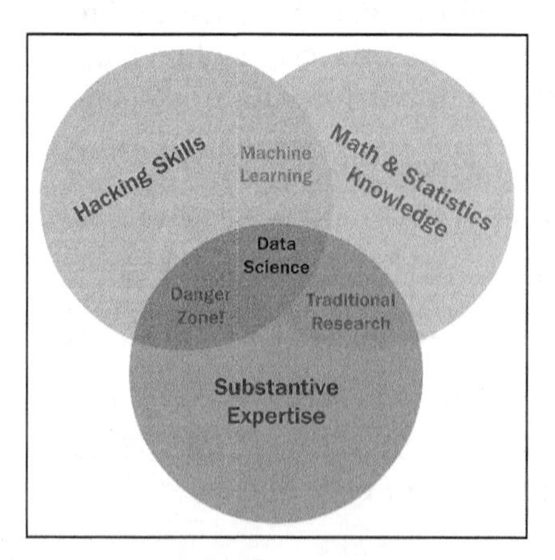

The Venn diagram of data science

Those with hacking skills can conceptualize and program complicated algorithms using computer languages. Having a **Math & Statistics Knowledge** base allows you to theorize and evaluate algorithms and tweak the existing procedures to fit specific situations. Having **Substantive Expertise** (domain expertise) allows you to apply concepts and results in a meaningful and effective way.

While having only two of these three qualities can make you intelligent, it will also leave a gap. Consider that you are very skilled in coding and have formal training in day trading. You might create an automated system to trade in your place but lack the math skills to evaluate your algorithms and, therefore, end up losing money in the long run. It is only when you can boast skills in coding, math, and domain knowledge that you can truly perform data science.

The one that was probably a surprise for you was **Domain Knowledge**. It is really just knowledge of the area you are working in. If a financial analyst started analyzing data about heart attacks, they might need the help of a cardiologist to make sense of a lot of the numbers.

Data Science is the intersection of the three key areas mentioned earlier. In order to gain knowledge from data, we must be able to utilize computer programming to access the data, understand the mathematics behind the models we derive, and above all, understand our analyses' place in the domain we are in. This includes the presentation of data. If we are creating a model to predict heart attacks in patients, is it better to create a PDF of information or an app where you can type in numbers and get a quick prediction? All these decisions must be made by the data scientist.

Also, note that the intersection of math and coding is machine learning. This book will look at machine learning in great detail later on but it is important to note that without the explicit ability to generalize any models or results to a domain, machine learning algorithms remain just that, algorithms sitting on your computer. You might have the best algorithm to predict cancer. You could be able to predict cancer with over 99% accuracy based on past cancer patient data but if you don't understand how to apply this model in a practical sense such that doctors and nurses can easily use it, your model might be useless.

Both computer programming and math are covered extensively in this book. Domain knowledge comes with both practice of data science and reading examples of other people's analyses.

The math

Most people stop listening once someone says the word *math*. They'll nod along in an attempt to hide their utter disdain for the topic. This book will guide you through the math needed for data science, specifically statistics and probability. We will use these subdomains of mathematics to create what are called **models**.

A **data model** refers to an organized and formal relationship between elements of data, usually meant to simulate a real-world phenomenon.

Essentially, we will use math in order to formalize relationships between variables. As a former pure mathematician and current math teacher, I know how difficult this can be. I will do my best to explain everything as clearly as I can. Between the three areas of data science, math is what allows us to move from domain to domain. Understanding the theory allows us to apply a model that we built for the fashion industry to a financial model.

The math covered in this book ranges from basic algebra to advanced probabilistic and statistical modeling. Do not skip over these chapters, even if you already know it or you're afraid of it. Every mathematical concept I introduce, I do so with care, examples, and purpose. The math in this book is essential for data scientists.

Example – spawner-recruit models

In biology, we use, among many others, a model known as the **spawner-recruit** model to judge the biological health of a species. It is a basic relationship between the number of healthy parental units of a species and the number of new units in the group of animals. In a public dataset of the number of salmon spawners and recruits, the following graph was formed to visualize the relationship between the two. We can see that there definitely is some sort of positive relationship (as one goes up, so does the other). But how can we formalize this relationship? For example, if we knew the number of spawners in a population, could we predict the number of recruits that group would obtain, and vice versa?

Essentially, models allow us to plug in one variable to get the other. Consider the following example:

$$Recruits = 0.5 * Spawners + 60$$

In this example, let's say we knew that a group of salmons had *1.15* (in thousands) of spawners. Then, we would have the following:

$$Recruits = 0.5 * 1.15 + 60$$

$$Recruits = 60.575(in\,thousands)$$

This result can be very beneficial to estimate how the health of a population is changing. If we can create these models, we can visually observe how the relationship between the two variables can change.

There are many types of data models, including probabilistic and statistical models. Both of these are subsets of a larger paradigm, called **machine learning**. The essential idea behind these three topics is that we use data in order to come up with the *best* model possible. We no longer rely on human instincts, rather, we rely on data.

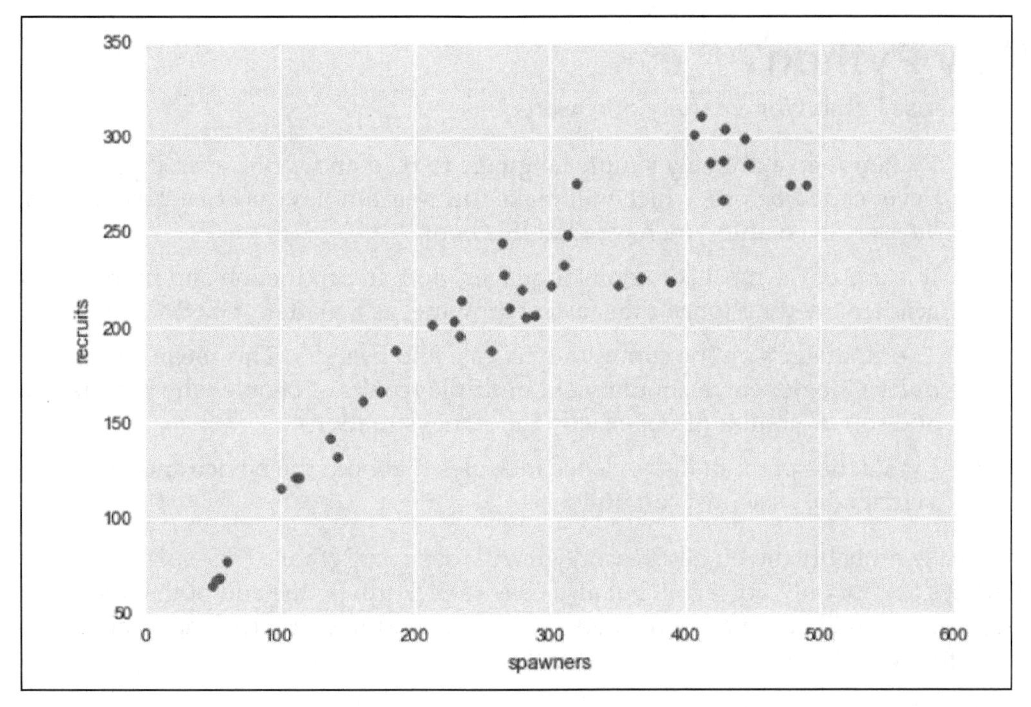

The spawner-recruit model visualized

The purpose of this example is to show how we can define relationships between data elements using mathematical equations. The fact that I used salmon health data was irrelevant! Throughout this book, we will look at relationships involving marketing dollars, sentiment data, restaurant reviews, and much more. The main reason for this is that I would like you (the reader) to be exposed to as many domains as possible.

Math and coding are vehicles that allow data scientists to step back and apply their skills virtually anywhere.

Computer programming

Let's be honest. You probably think computer science is way cooler than math. That's ok, I don't blame you. The news isn't filled with math news like it is with news on the technological front. You don't turn on the TV to see a new theory on primes, rather, you will see investigative reports on how the latest smartphone can take photos of cats better or something. Computer languages are how we communicate with the machine and tell it to do our bidding. A computer speaks many languages and, like a book, can be written in many languages; similarly, data science can also be done in many languages. Python, **Julia,** and **R** are some of the many languages available to us. This book will focus exclusively on using Python.

Why Python?

We will use Python for a variety of reasons:

- Python is an extremely simple language to read and write, even if you've never coded before, which will make future examples easy to ingest and read later on, even after you have read this book
- It is one of the most common languages, both in production and in the academic setting (one of the fastest growing, as a matter of fact)
- The language's online community is vast and friendly. This means that a quick Google search should yield multiple results of people who have faced and solved similar (if not exactly the same) situations
- Python has prebuilt data science modules that both the novice and the veteran data scientist can utilize

The last is probably the biggest reason we will focus on Python. These prebuilt modules are not only powerful, but also easy to pick up. By the end of the first few chapters, you will be very comfortable with these modules. Some of these modules are as follows:

- `pandas`
- `sci-kit learn`
- `seaborn`
- `numpy/scipy`
- `requests` (to mine data from the Web)
- `BeautifulSoup` (for the Web-HTML parsing)

Python practices

Before we move on, it is important to formalize many of the requisite coding skills in Python.

In Python, we have **variables** that are placeholders for objects. We will focus on only a few types of basic objects at first:

- `int` (an integer)
    - Examples: 3, 6, 99, -34, 34, 11111111
- `float` (a decimal):
    - Examples: 3.14159, 2.71, -0.34567
- `boolean` (either `True` or `False`)
    - The statement, Sunday is a weekend, is `True`
    - The statement, Friday is a weekend, is `False`
    - The statement, pi is exactly the ratio of a circle's circumference to its diameter, is `True` (crazy, right?)
- `string` (text or words made up of characters)
    - "I love hamburgers" (*by the way, who doesn't?*)
    - "Matt is awesome"
    - A Tweet is a string
- `list` (a collection of objects)
    - Example: [1, 5.4, True, "apple"]

We will also have to understand some basic logistical operators. For these operators, keep the `boolean` datatype in mind. Every operator will evaluate to either `True` or `False`. Let's take a look at the following illustrations:

- `==` evaluates to `True` if both sides are equal; otherwise it evaluates to `False`
    - 3 + 4 == 7 (will evaluate to `True`)
    - 3 - 2 == 7 (will evaluate to `False`)
- < (less than)
    - 3 < 5 (`True`)
    - 5 < 3 (`False`)

- <= (less than or equal to)
  - 3 <= 3 (`True`)
  - 5 <= 3 (`False`)
- > (greater than)
  - 3 > 5 (`False`)
  - 5 > 3 (`True`)
- >= (greater than or equal to)
  - 3 >= 3 (`True`)
  - 5 >= 3 (`False`)

When coding in Python, I will use a pound sign (#) to create a "comment," which will not be processed as code but is merely there to communicate with the reader. Anything to the right of a # sign is a comment on the code being executed.

Example of basic Python

In Python, we use spaces/tabs to denote operations that belong to other lines of code.

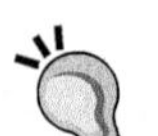

Note the use of the `if` statement. It means exactly what you think it means. When the statement after the `if` statement is `True`, then the tabbed part under it will be executed, as shown in the following code:

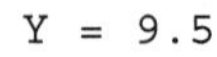

```
X = 5.8
Y = 9.5

X + Y == 15.3  # This is True!

X - Y == 15.3  # This is False!
5if x + y == 15.3:    # If the statement is true:
  print "True!"      # print something!
```

The `print "True!"` statement belongs to the `if x + y == 15.3:` line preceding it because it is tabbed right under it. This means that the `print` statement will be executed if and only if `x + y` equals `15.3`.

Note that the following `list` variable, `my_list`, can hold multiple types of objects. This one has an `int`, a `float`, `boolean`, and `string` inputs (in that order):

```
my_list = [1, 5.7, True, "apples"]

len(my_list) == 4  # 4 objects in the list

my_list[0] == 1    # the first object


my_list[1] == 5.7    # the second object
```

In the preceding code:

- I used the `len` command to get the length of the list (which was four).
- Note the zero-indexing of Python. Most computer languages start counting at zero instead of one. So if I want the first element, I call the index zero, and if I want the 95th element, I call the index 94.

Example – parsing a single tweet

Here is some more Python code. In this example, I will be parsing some tweets about stock prices (one of the important case studies in this book will be trying to predict market movements based on popular sentiment regarding stocks on social media):

```
tweet = "RT @j_o_n_dnger: $TWTR now top holding for
             Andor, unseating $AAPL"

words_in_tweet = first_tweet.split(' ') # list of words in tweet

for word in words_in_tweet:             # for each word in list
  if "$" in word:                       # if word has a "cashtag"
  print "THIS TWEET IS ABOUT", word  # alert the user
```

I will point out a few things about this code snippet, line by line, as follows:

- We set a variable to hold some text (known as a string in Python). In this example, the tweet in question is "`RT @robdv: $TWTR now top holding for Andor, unseating $AAPL`"
- The `words_in_tweet` variable *tokenizes* the tweet (separates it by word). If you were to print this variable, you would see the following:

```
['RT',
'@robdv:',
'$TWTR',
'now',
```

```
'top',
'holding',
'for',
'Andor,',
'unseating',
'$AAPL']
```

- We iterate through this list of words. This is called a `for` loop. It just means that we go through a list one by one.
- Here, we have another `if` statement. For each word in this tweet, if the word contains the `$` character (this is how people reference stock tickers on Twitter).
- If the preceding if statement is `true` (that is, if the tweet contains a cashtag), print it and show it to the user.

The output of this code will be as follows:

```
THIS TWEET IS ABOUT $TWTR
THIS TWEET IS ABOUT $AAPL
```

We get this output as these are the only words in the tweet that use the cashtag. Whenever I use Python in this book, I will ensure that I am as explicit as possible about what I am doing in each line of code.

Domain knowledge

As I mentioned earlier, this category focuses mainly on having knowledge about the particular topic you are working on. For example, if you are a financial analyst working on stock market data, you have a lot of domain knowledge. If you are a journalist looking at worldwide adoption rates, you might benefit from consulting an expert in the field. This book will attempt to show examples from several problem domains, including medicine, marketing, finance, and even UFO sightings!

Does that mean that if you're not a doctor, you can't work with medical data? Of course not! Great data scientists can apply their skills to any area, even if they aren't fluent in it. Data scientists can adapt to the field and contribute meaningfully when their analysis is complete.

A big part of domain knowledge is presentation. Depending on your audience, it can greatly matter how you present your findings. Your results are only as good as your vehicle of communication. You can predict the movement of the market with 99.99% accuracy, but if your program is impossible to execute, your results will go unused. Likewise, if your vehicle is inappropriate for the field, your results will go equally unused.

Some more terminology

This is a good time to define some more vocabulary. By this point, you're probably excitedly looking up a lot of data science material and seeing words and phrases I haven't used yet. Here are some common terminologies you are likely to come across:

- **Machine learning**: This refers to giving computers the ability to learn from data without explicit "rules" being given by a programmer.

  We have seen the concept of machine learning earlier in this chapter as the union of someone who has both coding and math skills. Here, we are attempting to formalize this definition. Machine learning combines the power of computers with intelligent learning algorithms in order to automate the discovery of relationships in data and create of powerful data models. Speaking of data models, we will concern ourselves with the following two basic types of data models:

- **Probabilistic model**: This refers to using probability to find a relationship between elements that includes a degree of randomness.
- **Statistical model**: This refers to taking advantage of statistical theorems to formalize relationships between data elements in a (usually) simple mathematical formula.

While both the statistical and probabilistic models can be run on computers and might be considered machine learning in that regard, we will keep these definitions separate as machine learning algorithms generally attempt to *learn* relationships in different ways.

We will take a look at the statistical and probabilistic models in the later chapters.

- **Exploratory data analysis (EDA)** refers to preparing data in order to standardize results and gain quick insights.

  EDA is concerned with data visualization and preparation. This is where we turn unorganized data into organized data and also clean up missing/incorrect data points. During EDA, we will create many types of plots and use these plots to identify key features and relationships to exploit in our data models.

- **Data mining** is the process of finding relationships between elements of data.

  Data mining is the part of data science where we try to find relationships between variables (think spawn-recruit model).

- I tried pretty hard not to use the term **big data** up until now. This is because I think this term is misused, a lot. While the definition of this word varies from person, big data. Big Data is data that is too large to be processed by a single machine (if your laptop crashed, it might be suffering from a case of big data).

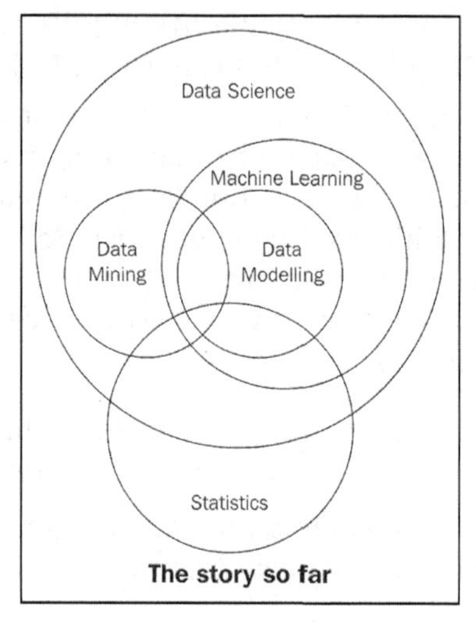

The state of data science (so far). This diagram is incomplete and is meant for visualization purposes only.

Data science case studies

The combination of math, computer programming, and domain knowledge is what makes data science so powerful. Often, it is difficult for a single person to master all three of these areas. That's why it's very common for companies to hire teams of data scientists instead of a single person. Let's look at a few powerful examples of data science in action and their outcome.

Case study – automating government paper pushing

Social security claims are known to be a major hassle for both the agent reading it and for the person who wrote the claim. Some claims take over 2 years to get resolved in their entirety, and that's absurd! Let's look at what goes into a claim:

B. To be completed by the claimant

PLEASE PRINT

Please Answer the Following Questions:

(1) Have you been treated or examined by a doctor (other than a doctor at a hospital) since the above date? ☐ Yes ☐ No

(If yes, please list the names, addresses and telephone numbers of doctors who have treated or examined you since the above date. Also list the dates of treatment or examination. If possible, send updated reports from these doctors to the Administrative Law Judge before the date of your hearing.)

DOCTORS NAME(S)	ADDRESS(ES) & TELEPHONE NO.(S)	DATE(S)

(2) What have these doctors told you about your condition?

(3) Have you been hospitalized since the above date? ☐ Yes ☐ No

(If yes, please list the name and address of the hospital. Also, explain why you were hospitalized and what treatment you received.)

Sample social security form

Not bad. It's mostly just text, though. Fill this in, then that, then this, and so on. You can see how it would be difficult for an agent to read these all day, form after form. There must be a better way!

Well, there is. Elder Research Inc. parsed this unorganized data and was able to automate 20% of all disability social security forms. This means that a computer could look at 20% of these written forms and give its opinion on the approval.

Not only that, the third-party company that is hired to rate the approvals of the forms actually gave the machine-graded forms a higher grade than the human forms. So, not only did the computer handle 20% of the load, it, on average, did better than a human.

Fire all humans, right?

Before I get a load of angry e-mails claiming that data science is bringing about the end of human workers, keep in mind that the computer was *only* able to handle 20% of the load. That means it probably performed terribly for 80% of the forms! This is because the computer was probably great at *simple forms*. The claims that would have taken a human minutes took the computer seconds to compute. But these minutes add up, and before you know it, each human is being saved over an hour a day!

Forms that might be easy for a human to read are also likely easy for the computer. It's when the form becomes very terse or when the writer starts deviating from usual grammar that the computer starts to fail. This model is great because it lets the humans spend more time on those difficult claims and gives them more attention without getting distracted by the sheer volume of papers.

Note that I used the word *model*. Remember that a model is a relationship between elements. In this case, the relationship is between written words and the approval status of a claim.

Case study – marketing dollars

A dataset shows the relationship between the money spent in the categories of TV, radio, and newspaper. The goal is to analyze the relationship between the three different marketing mediums and how it affects the sale of a product. Our data is in the form of a row and column structure. Each row represents a sales region and the columns tell us how much money was spent on each medium and the profit achieved in that region.

Usually, the data scientist must ask for units and scale. In this case, I will tell you that TV, radio, and newspaper are measured in "thousands of dollars" and sales in "thousands of widgets sold". This means that in the first region, $230,100 was spent on TV advertising, $37,800 on radio advertising, and $69,200 on newspaper advertising. In the same region, 22,100 items were sold.

	TV	Radio	Newspaper	Sales
1	230.1	37.8	69.2	22.1
2	44.5	39.3	45.1	10.4
3	17.2	45.9	69.3	9.3
4	151.5	41.3	58.5	18.5
5	180.8	10.8	58.4	12.9

Advertising budgets

For example, in the third region, we spent $17,200 on TV advertising and sold 9,300 widgets.

If we plot each variable against sales, we get the following graphs:

```
import seaborn as sns
sns.pairplot(data, x_vars=['TV','Radio','Newspaper'], y_vars='Sales')
```

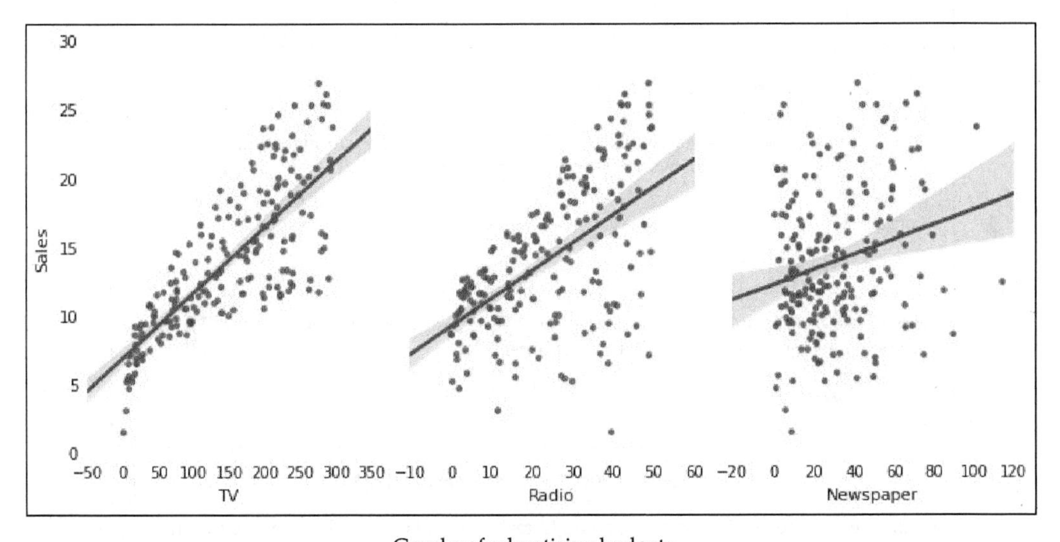

Graphs of advertising budgets

Note how none of these variables form a very strong line and, therefore, might not work well to predict sales (on their own). TV comes closest in forming an obvious relationship, but still even that isn't great. In this case, we will have to form a more complex model than the one we used in the spawner-recruiter model and combine all three variables in order to model sales.

This type of problem is very common in data science. In this example, we are attempting to identify key features that are associated with the sales of a product. If we can isolate these key features, then we can exploit these relationships and change how much we spend on advertising in different places with the hopes of increasing our sales.

Case study – what's in a job description?

Looking for a job in data science? Great, let me help. In this case study, I have "scraped" (taken from the Web) 1,000 job descriptions for companies actively hiring data scientists (as of January 2016). The goal here is to look at some of the most common keywords people use in their job descriptions.

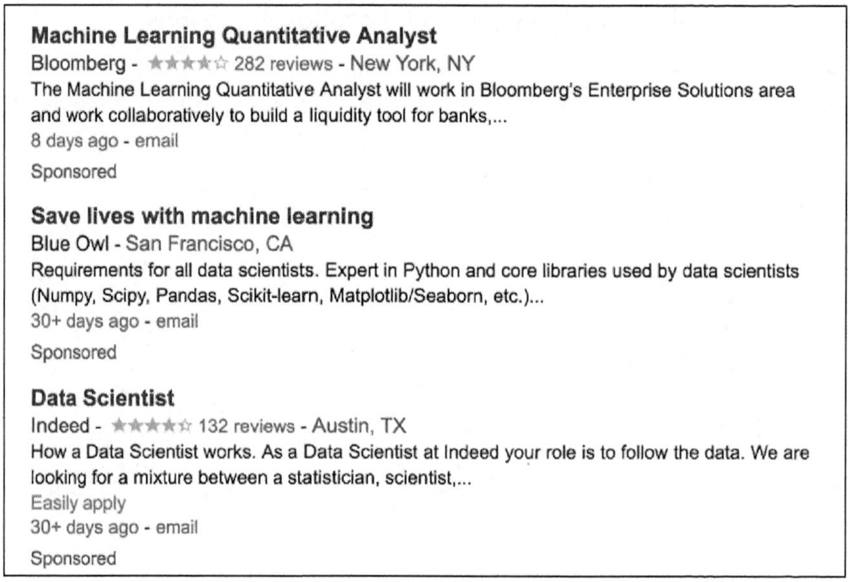

Machine Learning Quantitative Analyst
Bloomberg - ★★★★☆ 282 reviews - New York, NY
The Machine Learning Quantitative Analyst will work in Bloomberg's Enterprise Solutions area and work collaboratively to build a liquidity tool for banks,...
8 days ago - email
Sponsored

Save lives with machine learning
Blue Owl - San Francisco, CA
Requirements for all data scientists. Expert in Python and core libraries used by data scientists (Numpy, Scipy, Pandas, Scikit-learn, Matplotlib/Seaborn, etc.)...
30+ days ago - email
Sponsored

Data Scientist
Indeed - ★★★★☆ 132 reviews - Austin, TX
How a Data Scientist works. As a Data Scientist at Indeed your role is to follow the data. We are looking for a mixture between a statistician, scientist,...
Easily apply
30+ days ago - email
Sponsored

An example of data scientist job listings.

(Note the second one asking for core Python libraries; we talk about these later on in this book)

```
import requests
# used to grab data from the web

from BeautifulSoup import BeautifulSoup
# used to parse HTML

from sklearn.feature_extraction.text import CountVectorizer
# used to count number of words and phrases (we will be using this
module a lot)
```

The first two imports are used to grab web data from the website, `Indeed.com`, and the third import is meant to simply count the number of times a word or phrase appears.

```
texts = []
# hold our job descriptions in this list

for index in range(0,1000,10): # go through 100 pages of indeed
  page = 'indeed.com/jobs?q=data+scientist&start='+str(index)
  # identify the url of the job listings

  web_result = requests.get(page).text
  # use requests to actually visit the url

  soup  BeautifulSoup(web_result)
  # parse the html of the resulting page

  for listing in soup.findAll('span', {'class':'summary'}:
    # for each listing on the page

    texts.append(listing.text)
 # append the text of the listing to our list
```

Okay, before I lose you, all that this loop is doing is going through 100 pages of job descriptions, and for each page, grabbing each job description. The important variable here is `texts`, which is a list of over 1,000 job descriptions:

```
type(texts) # == list

vect = CountVectorizer(ngram_range=(1,2), stop_words='english')
# Get basic counts of one and two word phrases

matrix = vect.fit_transform(texts)
# fit and learn to the vocabulary in the corpus

print len(vect.get_feature_names())  # how many features are there
# There are 11,293 total one and two words phrases in my case!!
```

I have omitted some code here, but it exists in the GitHub repository for this book. The results are as follows (represented as the phrase, and then the number of of times it occurred):

```
experience 320
machine 306
learning 305
machine learning 294
techniques 266
```

```
statistical 215
team 197
analytics 173
business 167
statistics 159
algorithms 152
datamining 149
software 144
applied 141
programming 132
understanding 127
world 127
research 125
datascience 123
methods 122
join 122
quantitative 122
group 121
real 120
large 120
```

Notable things:

- Machine learning and experience are at the top of the list. Experience comes with practice. A basic idea of machine learning comes with this book.
- These words are followed closely by statistical words implying knowledge of math and theory.
- The word `team` is very high up, implying that you will need to work with a team of data scientists; you won't be a lone wolf.
- Computer science words such as `algorithms` and `programming` are prevalent.
- The words `techniques`, `understanding`, and `methods` imply a more theoretical approach, ambivalent to any single domain.
- The word `business` implies a particular problem domain.

There are many interesting things to note about this case study but the biggest take away is that there are many key words and phrases that make up a data science role. It isn't just math, coding, or domain knowledge; it truly is the combination of these three ideas (whether exemplified in a single person or across a multiperson team) that makes data science possible and powerful.

Summary

At the beginning of this chapter, I posed a simple question, what's the catch of data science? Well there is one. It isn't all fun, games and modelling. There must be a price to our quest for ever smarter machines and algorithms. As we seek new and innovative ways to discover data trends, a beast lurks in the shadows. I'm not talking about the learning curve of mathematics or programming nor am I referring to the surplus of data. The industrial age left us with an ongoing battle against pollution. The subsequent information age left behind a trail of big data. So, what dangers might the data age bring us?

The data age can lead to something much more sinister—the dehumanization of the individual through mass data.

More and more people are jumping headfirst into the field of data science, most with no prior experience in math or CS, which on the surface is great. Average data scientists have access to millions of dating profiles' data, tweets, online reviews, and much more in order to jumpstart their education.

However, if you jump into data science without the proper exposure to theory or coding practices and without respect of the domain you are working in, you face the risk of oversimplifying the very phenomenon you are trying to model.

For example, let's say you want to automate your sales pipeline by building a simplistic program that looks at LinkedIn for very specific keywords in a person's LinkedIn profile.

```
keywords = ["Saas", "Sales", "Enterprise"]
```

Great, now you can scan LinkedIn quickly to find people who match your criteria. But what about that person who spells out "Software as a Service" instead of "Saas" or misspells "enterprise" (it happens to the best of us; I bet someone will find a typo in my book). How will your model figure out that these people are also a good match? They should not be left behind just because the cut corners data scientist has overgeneralized people in such an easy way.

The programmer chose to simplify their search for another human by looking for three basic keywords and ended up with a lot of missed opportunities left on the table.

In the next chapter, we will explore the different types of data that exist in the world, ranging from free-form text to highly structured row/column files. We will also look at the mathematical operations that are allowed for different types of data, as well as deduce insights based on the form the data that comes in.

2
Types of Data

Now that we have a basic introduction to the world of data science and understand why the field is so important, let's take a look at the various ways in which data can be formed. Specifically, in this chapter we will look at the following topics:

- Structured versus unstructured data
- Quantitative versus qualitative data
- The four levels of data

We will dive further into each of these topics by showing examples of how data scientists look at and work with data. This chapter is aimed to familiarize ourselves with the fundamental ideas underlying data science.

Flavors of data

In the field, it is important to understand the different flavors of data for several reasons. Not only will the type of data dictate the methods used to analyze and extract results, knowing whether the data is unstructured or perhaps quantitative can also tell you a lot about the real-world phenomenon being measured.

We will look at the three basic classifications of data:

- Structured vs unstructured (sometimes called organized vs unorganized)
- Quantitative vs qualitative
- The four levels of data

The first thing to pay attention to is my use of the word *data*. In the last chapter, I defined data as merely being a collection of information. This vague definition exists because we may separate data into different categories and need our definition to be loose.

The next thing to remember while we go through this chapter is that for the most part, when I talk about what type of *data* this is, I will refer to either a *specific characteristic* of a dataset or to the *entire dataset* as a whole. I will be very clear about which one I refer to at any given time.

Why look at these distinctions?

It might seem worthless to stop and think about what type of data we have before getting into the fun stuff, like statistics and machine learning, but this is arguably one of the most important steps you need to take to perform data science.

Consider an example where we are looking at election results for a county. In the dataset of people, there is a "race" column that is denoted via an identifying number to save space. For example perhaps caucasian is denoted by 7 while Asian American is 2. Without understanding that these numbers are not actually ordered numbers like we think about them (where 7 is greater than 2 and therefore caucasian is "greater than" Asian American) we will make terrible mistakes in our analysis. Discuss

The same principle applies to data science. When given a dataset, it is tempting to jump right into exploring, applying statistical models, and researching the applications of machine learning in order to get results faster. However, if you don't understand the type of data that you are working with, then you might waste a lot of time applying models that are known to be ineffective with that specific type of data.

When given a new dataset, I always recommend taking about an hour (usually less) to make the distinctions mentioned in the following sections.

Structured versus unstructured data

The distinction between structured and unstructured data is usually the first question you want to ask yourself about the *entire* dataset. The answer to this question can mean the difference between needing three days or three weeks of time to perform a proper analysis.

The basic breakdown is as follows (this is a rehashed definition of organized and unorganized data in the first chapter):

- **Structured (organized) data**: This is data that can be thought of as observations and characteristics. It is usually organized using a table method (rows and columns).
- **Unstructured (unorganized) data**: This data exists as a free entity and does not follow any standard organization hierarchy.

Here are a few examples that could help you differentiate between the two:

- Most data that exists in text form, including server logs and Facebook posts, is *unstructured*
- Scientific observations, as recorded by careful scientists, are kept in a very neat and organized (*structured*) format
- A genetic sequence of chemical nucleotides (for example, ACGTATTGCA) is *unstructured* even if the order of the nucleotides matters as we cannot form descriptors of the sequence using a row/column format without taking a further look

Structured data is generally thought of as being much easier to work with and analyze. Most statistical and machine learning models were built with structured data in mind and cannot work on the loose interpretation of unstructured data. The natural row and column structure is easy to digest for human and machine eyes. So why even talk about unstructured data? Because it is so common! Most estimates place unstructured data as 80-90% of the world's data. This data exists in many forms and for the most part, goes unnoticed by humans as a potential source of data. Tweets, e-mails, literature, and server logs are generally unstructured forms of data.

While a data scientist likely prefers structured data, they must be able to deal with the world's massive amounts of unstructured data. If 90% of the world's data is unstructured, that implies that about 90% of the world's information is trapped in a difficult format.

So, with most of our data existing in this free-form format, we must turn to pre-analysis techniques, called *preprocessing*, in order to apply structure to at least a part of the data for further analysis. The next chapter will deal with preprocessing in great detail; for now, we will consider the part of preprocessing wherein we attempt to apply transformations to convert unstructured data into a structured counterpart.

Example of data preprocessing

When looking at text data (which is almost always considered unstructured), we have many options to transform the set into a structured format. We may do this by applying new characteristics that describe the data. A few such characteristics are as follows:

- Word/phrase count
- The existence of certain special characters
- The relative length of text
- Picking out topics

I will use the following tweet as a quick example of unstructured data, but you may use any unstructured free-form text that you like, including tweets and Facebook posts.

This Wednesday morn, are you early to rise? Then look East. The Crescent Moon joins Venus & Saturn. Afloat in the dawn skies.

It is important to reiterate that pre-processing is necessary for this tweet because a vast majority of learning algorithms require numerical data (which we will get into after this example).

More than requiring a certain type of data, pre-processing allows us to explore features that have been created from the existing features. For example, we can extract features such as word count and special characters from the mentioned tweet. Now, let's take a look at a few features that we can extract from text.

Word/phrase counts

We may break down a tweet into its word/phrase count. The word *this* appears in the tweet once, as does every other word. We can represent this tweet in a structured format, as follows, thereby converting the unstructured set of words into a row/column format:

	this	**wednesday**	**morn**	**are**	**this wednesday**
Word Count	1	1	1	1	1

Note that to obtain this format we can utilize scikit-learn's `CountVectorizer` that we saw in the previous chapter.

Presence of certain special characters

We may also look at the presence of special characters, such as the question mark and exclamation mark. The appearance of these characters might imply certain ideas about the data that are otherwise difficult to know. For example, the fact that this tweet contains a question mark might strongly imply that this tweet contains a question for the reader. We might append the preceding table with a new column, as shown:

	this	**wednesday**	**morn**	**are**	**this wednesday**	**?**
Word Count	1	1	1	1	1	1

Relative length of text

This tweet is 121 characters long.

```
len("This Wednesday morn, are you early to rise? Then look East. The
Crescent Moon joins Venus & Saturn. Afloat in the dawn skies.")
# get the length of this text (number of characters for a string)

# 121
```

The average tweet, as discovered by analysts, is about 30 characters in length. So, we might impose a new characteristic, called **relative length**, (which is the length of the tweet divided by the average length), telling us the length of this tweet as compared to the average tweet. This tweet is actually 4.03 times longer than the average tweet, as shown:

$$\frac{121}{30} = 4.03$$

We can add yet another column to our table using this method:

	this	wednesday	morn	are	this wednesday	?	Relative length
Word Count	1	1	1	1	1	1	4.03

Picking out topics

We can pick out some topics of the tweet to add as columns. This tweet is about astronomy, so we can add another column, as illustrated:

	this	wednesday	morn	are	this wednesday	?	Relative length	Topic
Word Count	1	1	1	1	1	1	4.03	astronomy

And just like that, we can convert a piece of text into structured/organized data ready for use in our models and exploratory analysis.

Topic is the only extracted feature we looked at that is not automatically derivable from the tweet. Looking at word count and tweet length in Python is easy; however, more advanced models (called topic models) are able to derive and predict topics of natural text as well.

Being able to quickly recognize whether your data is structured or unstructured can save hours or even days of work in the future. Once you are able to discern the organization of the data presented to you, the next question is aimed at the individual characteristics of the dataset.

Quantitative versus qualitative data

When you ask a data scientist, "what type of data is this?", they will usually assume that you are asking them whether or not it is mostly quantitative or qualitative. It is likely the most common way of describing the *specific* characteristics of a dataset.

For the most part, when talking about quantitative data, you are *usually* (not always) talking about a structured dataset with a strict row/column structure (because we don't assume unstructured data even *has* any characteristics). All the more reason why the preprocessing step is so important.

These two data types can be defined as follows:

- **Quantitative data**: This data can be described using numbers, and basic mathematical procedures, including addition, are possible on the set.
- **Qualitative data**: This data cannot be described using numbers and basic mathematics. This data is generally thought of as being described using "natural" categories and language.

Example – coffee shop data

Say that we were processing observations of coffee shops in a major city using the following five descriptors (characteristics):

Data: Coffee Shop

- Name of coffee shop
- Revenue (in thousands of dollars)
- Zip code
- Average monthly customers
- Country of coffee origin

Each of these characteristics can be classified as either quantitative or qualitative, and that simple distinction can change everything. Let's take a look at each one:

- Name of coffee shop - Qualitative

  The name of a coffee shop is not expressed as a number and we cannot perform math on the name of the shop.

- Revenue - Quantitative

  How much money a cafe brings in can definitely be described using a number. Also, we can do basic operations such as adding up the revenue for 12 months to get a year's worth of revenue.

- Zip code - Qualitative

  This one is tricky. A zip code is always represented using numbers, but what makes it qualitative is that it does not fit the second part of the definition of quantitative—we cannot perform basic mathematical operations on a zip code. If we add together two zip codes, it is a nonsensical measurement. We don't necessarily get a new zip code and we definitely don't get "double the zip code".

- Average monthly customers - Quantitative

  Again, describing this factor using numbers and addition makes sense. Add up all of your monthly customers and you get your yearly customers.

- Country of coffee origin - Qualitative

  We will assume this is a very small café with coffee from a single origin. This country is described using a name (Ethiopian, Colombian), and not numbers.

A couple of important things to note:

- Even though a zip code is being described using numbers, it is not quantitative. This is because you can't talk about the *sum* of all zip codes or an *average* zip code. These are nonsensical descriptions.
- Pretty much whenever a word is used to describe a characteristic, it is a qualitative factor.

If you are having trouble identifying which is which, basically, when trying to decide whether or not the data is qualitative or quantitative, ask yourself a few basic questions about the data characteristics:

- Can you describe it using numbers?
  - No? It is **qualitative.**
  - Yes? Move on to next question.

- Does it still makes sense after you add them together?
    - No? They are **qualitative.**
    - Yes? You probably have **quantitative** data.

This method will help you classify most, if not all, data into one of these two categories.

The difference between these two categories define the types of questions you may ask about each column. For a quantitative column, you may ask questions such as the following:

- What is the average value?
- Does this quantity increase or decrease over time (if time is a factor)?
- Is there a threshold that if this number grew above or be too low would signal trouble for the company?

For a qualitative column, none of the preceding questions can be answered; however, the following questions *only* apply to qualitative values:

- Which value occurs the most and the least?
- How many unique values are there?
- What are these unique values?

Example – world alcohol consumption data

The World Health Organization released a dataset describing the average drinking habits of people in countries across the world. We will use Python and the data exploration tool, Pandas, in order to gain a better look:

```
import pandas as pd

# read in the CSV file from a URL
drinks = pd.read_csv('https://raw.githubusercontent.com/sinanuozdemir/
principles_of_data_science/master/data/chapter_2/drinks.csv')

# examine the data's first five rows
drinks.head()            # print the first 5 rows
```

These three lines have done the following:

- Imported `pandas`, which will be referred to as `pd` in the future
- Read in a **CSV** (**comma separated value**) file as a variable called `drinks`
- Called a method, `head`, that reveals the first five rows of the dataset

Note the neat row/column structure a CSV comes in

	country	beer_servings	spirit_servings	wine_servings	total_litres_of_pure_alcohol	continent
0	Afghanistan	0	0	0	0.0	AS
1	Albania	89	132	54	4.9	EU
2	Algeria	25	0	14	0.7	AF
3	Andorra	245	138	312	12.4	EU
4	Angola	217	57	45	5.9	AF

We have six different columns that we are working with in this example:

- `country`: Qualitative
- `beer_servings`: Quantitative
- `spirit_servings`: Quantitative
- `wine_servings`: Quantitative
- `total_litres_of_pure_alcohol`: Quantitative
- `continent`: Qualitative

Let's look at the qualitative column `continent`. We can use Pandas in order to get some basic summary statistics about this non-numerical characteristic. The `describe()` method is being used here, which first identifies whether the column is likely quantitative or qualitative and then gives basic information about the column as a whole. This is shown as follows:

```
drinks['continent'].describe()

>> count      193
>> unique       5
>> top         AF
>> freq        53
```

It reveals that the WHO has gathered data about five unique continents, the most frequent being `AF` (Africa), which occurred 53 times in the 193 observations.

If we take a look at one of the quantitative columns and call the same method, we can see the difference in output, as shown:

```
drinks['beer_servings'].describe()

>> mean      106.160622
>> min         0.000000
>> max       376.000000
```

Now we can look at the mean (average) beer serving per-person per-country (106.2 servings) as well as the lowest beer serving, zero, and the highest beer serving recorded, 376 (that's more than a beer a day).

Digging deeper

Quantitative data can be broken down, one step further, into *discrete* and *continuous* quantities.

These can be defined as follows:

- **Discrete data**: This describes data that is counted. It can only take on certain values.

  Examples of discrete quantitative data include a dice roll, because it can only take on six values, and the number of customers in a café, because you can't have a *real* range of people.

- **Continuous data**: This describes data that is measured. It exists on an infinite range of values.

  A good example of continuous data would be a person's weight because it can be 150 pounds or 197.66 pounds (note the decimals). The height of a person or building is a continuous number because an infinite scale of decimals is possible. Other examples of continuous data would be time and temperature.

The road thus far...

So far in this chapter, we have looked at the differences between structured and unstructured data as well as between qualitative and quantitative characteristics. These two simple distinctions can have drastic effects on the analysis that is performed. Allow me to summarize before moving on the second half of the chapter.

Data as a whole can either be *structured* or *unstructured*, meaning that the data can either take on an organized row/column structure with distinct features that describe each row of the dataset, or exist in a free-form state that usually must be preprocessed into a form that is easily digestible.

If data is structured, we can look at each column (feature) of the dataset as being either *quantitative* or *qualitative*. Basically, can the column be described using mathematics and numbers or not? The next part of this chapter will break down data into four very specific and detailed levels. At each order, we will apply more complicated rules of mathematics, and in turn, we can gain a more intuitive and quantifiable understanding of the data.

The four levels of data

It is generally understood that a specific characteristic (feature/column) of structured data can be broken down into one of four levels of data. The levels are:

- The nominal level
- The ordinal level
- The interval level
- The ratio level

As we move down the list, we gain more structure and, therefore, more returns from our analysis. Each level comes with its own accepted practice in measuring the `center` of the data. We usually think of the mean/average as being an acceptable form of center, however, this is only true for a specific type of data.

The nominal level

The first level of data, the *nominal* level, (which also sounds like the word name) consists of data that is described purely by name or category. Basic examples include gender, nationality, species, or yeast strain in a beer. They are not described by numbers and are therefore qualitative. The following are some examples:

- A type of animal is on the nominal level of data. We may also say that if you are a chimpanzee, then you belong to the mammalian class as well.
- A part of speech is also considered on the nominal level of data. The word *she* is a pronoun, and it is also a *noun*.

Of course, being qualitative, we cannot perform any quantitative mathematical operations, such as addition or division. These would not make any sense.

Mathematical operations allowed

We cannot perform mathematics on the nominal level of data except the basic *equality* and *set membership* functions, as shown in the following two examples:

- *Being a tech entrepreneur* is the same as *being in the tech industry*, but not vice versa
- A figure described as a square falls under the description of being a rectangle, but not vice versa

Measures of center

A **measure of center** is a number that describes what the data *tends to*. It is sometimes referred to as the *balance point* of the data. Common examples include the mean, median, and mode.

In order to find the *center* of nominal data, we generally turn to the *mode* (the most common element) of the dataset. For example, look back at the WHO alcohol consumption data. The most common continent surveyed was Africa, making that a possible choice for the *center* of the continent column.

Measures of center such as the mean and median do not make sense at this level as we cannot order the observations or even add them together.

What data is like at the nominal level

Data at the nominal level is mostly categorical in nature. Because we generally can only use words to describe the data, it can be lost in translation among countries, or can even be misspelled.

While data at this level can certainly be useful, we must be careful about what insights we may draw from them. With only the mode as a basic measure of center, we are unable to draw conclusions about an *average* observation. This concept does not exist at this level. It is only at the next level that we may begin to perform true mathematics on our observations.

The ordinal level

The nominal level did not provide us with much flexibility in terms of mathematical operations due to one seemingly unimportant fact—we could not order the observations in any natural way. Data in the *ordinal* level provides us with a rank order, or the means to place one observation before the other; however, it does not provide us with relative differences between observations, meaning that while we may order the observations from first to last, we cannot add or subtract them to get any real meaning.

Examples

The *Likert* is among the most common ordinal level scales. Whenever you are given a survey asking you to rate your satisfaction on a scale from 1 to 10, you are providing data at the ordinal level. Your answer, which must fall between 1 and 10, can be ordered: eight is better than seven while three is worse than nine.

However, differences between the numbers do not make much sense. The difference between a seven and a six might be different than the difference between a two and a one.

Mathematical operations allowed

We are allowed much more freedom on this level in mathematical operations. We inherit all mathematics from the ordinal level (equality and set membership) and we can also add the following to the list of operations allowed in the nominal level:

- Ordering
- Comparison

Ordering refers to the natural order provided to us by the data; however, this can be tricky to figure out sometimes. When speaking about the spectrum of visible light, we can refer to the names of colors—red, orange, yellow, green, blue, indigo, and violet. Naturally, as we move from left to right, the light is gaining energy and other properties. We may refer to this as a natural order.

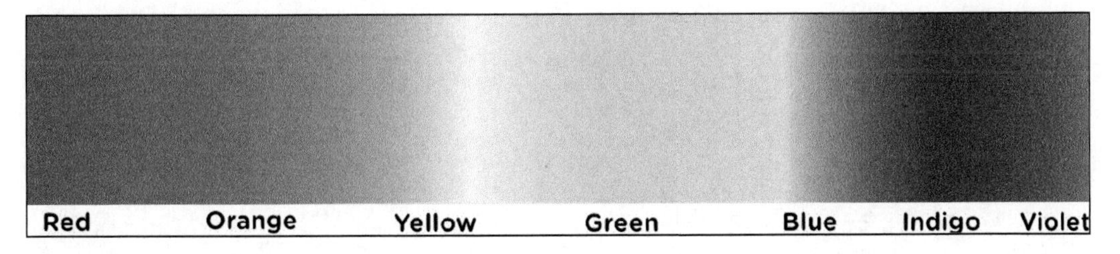

However, if needed, an artist may impose another order on the data, such as sorting the colors based on the cost of the material to make the said color. This could change the order of the data but as long as we are consistent in what defines the order, it does not matter what defines it.

Comparisons are another new operation allowed at this level. At the ordinal level, it would not make sense to say that one country was *naturally* better than another or that one part of speech is worse than another. At the ordinal level, we can make these comparisons. For example, we can talk about how putting a "7" on a survey is worse than putting a "10".

Measures of center

At the ordinal level, the **median** is usually an appropriate way of defining the center of the data. The mean, however, would be impossible because division is not allowed at this level. We can also use the mode like we could at the nominal level.

We will now look at an example of using the median:

Imagine you have conducted a survey among your employees asking "how happy are you to be working here on a scale from 1-5", and your results are as follows:

```
5, 4, 3, 4, 5, 3, 2, 5, 3, 2, 1, 4, 5, 3, 4, 4, 5, 4, 2, 1, 4, 5, 4,
3, 2, 4, 4, 5, 4, 3, 2, 1
```

Let's use Python to find the median of this data. It is worth noting that most people would argue that the mean of these scores would work just fine. The reason that the mean would not be as mathematically viable is because if we subtract/add two scores, say a score of four minus a score of two, the difference of two does not really mean anything. If addition/subtraction among the scores doesn't make sense, the mean won't make sense either.

```
import numpy

results = [5, 4, 3, 4, 5, 3, 2, 5, 3, 2, 1, 4, 5, 3, 4, 4, 5, 4, 2, 1,
4, 5, 4, 3, 2, 4, 4, 5, 4, 3, 2, 1]


sorted_results = sorted(results)


print sorted_results
'''
[1, 1, 1, 2, 2, 2, 2, 2, 3, 3, 3, 3, 3, 3, 4, 4, 4, 4, 4, 4, 4, 4, 4,
4, 4, 5, 5, 5, 5, 5, 5, 5]
'''


print numpy.mean(results)     # == 3.4375


print numpy.median(results)  # == 4.0
```

The ''' (triple apostrophe) denotes a longer (over two lines) comment. It acts in a way similar to the #.

Turns out that the median is not only more sound, but makes the survey results look much better.

Quick recap and check

So far we have seen half of the levels of data:

- The nominal level
- The ordinal level

At the nominal level, we deal with data usually described using vocabulary (but sometimes with numbers), with no order, and little use of mathematics.

At the ordinal level, we have data that can be described with numbers and also have a "natural" order, allowing us to put one in front of the other.

Let's try to classify the following example as either ordinal or nominal (answers are at the end of the chapter):

- The origin of the beans in your cup of coffee
- The place someone receives after completing a foot race
- The metal used to make the medal that they receive after placing in the said race
- The telephone number of a client
- How many cups of coffee you drink in a day

The interval level

Now we are getting somewhere interesting. At the interval level, we are beginning to look at data that can be expressed through very quantifiable means, and where much more complicated mathematical formulas are allowed. The basic difference between the ordinal level and the interval level is, well, just that—difference.

Data at the interval level allows meaningful subtraction between data points.

Example

Temperature is a great example of data at the interval level. If it is 100 degrees Fahrenheit in Texas and 80 degrees Fahrenheit in Istanbul, Turkey, then Texas is 20 degrees warmer than Istanbul. This simple example allows for so much more manipulation at this level than previous examples.

(Non) Example

It seems as though the example in the ordinal level (using the one to five survey) fits the bill of the interval level. However, remember that the *difference* between the scores (when you subtract them) does not make sense, therefore, this data cannot be called at the interval level.

Mathematical operations allowed

We can use all the operations allowed on the lower levels (ordering, comparisons, and so on), alongwith two other notable operations:

- Addition
- Subtraction

The allowance of these two operations allows us to talk about data at this level in a whole new way.

Measures of center

At this level, we can use the median and mode to describe this data; however, usually the most accurate description of the center of data would be the **arithmetic mean**, more commonly referred to as, simply, "the mean". Recall that the definition of the mean requires us to add together all the measurements. At the previous levels, addition was meaningless; therefore, the mean would have lost extreme value. It is only at the interval level and above that the arithmetic mean makes sense.

We will now look at an example of using the mean.

Suppose we look at the temperature of a fridge containing a pharmaceutical company's new vaccine. We measure the temperate every hour with the following data points (in Fahrenheit):

```
31, 32, 32, 31, 28, 29, 31, 38, 32, 31, 30, 29, 30, 31, 26
```

Using Python again, let's find the mean and median of the data:

```
import numpy


temps = [31, 32, 32, 31, 28, 29, 31, 38, 32, 31, 30, 29, 30, 31, 26]


print numpy.mean(temps)     # == 30.73

print numpy.median(temps)   # == 31.0
```

Note how the mean and median are quite close to each other and both are around 31 degrees. The question, *on average, how cold is the fridge?*, about 31, however the vaccine comes with a warning:

Do not keep this vaccine at a temperature under 29 degrees.

Note that at least twice, the temperature dropped below 29 degrees but you ended up assuming that it isn't enough for it to be detrimental.

This is where the measure of variation can help us understand how bad the fridge situation can be.

Measures of variation

This is something new that we have not yet discussed. It is one thing to talk about the center of the data but, in data science, it is also very important to mention how "spread out" the data is. The measures that describe this phenomenon are called **measures of variation**. You have likely heard of "standard deviation" before and are now experiencing mild PTSD from your statistics classes. This idea is extremely important and I would like to address it briefly.

A measure of variation (like the standard deviation) is a number that attempts to describe how spread out the data is.

Along with a measure of center, a measure of variation can almost entirely describe a dataset with only two numbers.

Standard deviation

Arguably, standard deviation is the most common measure of variation of data at the interval level and beyond. The standard deviation can be thought of as the "average distance a data point is at from the mean". While this description is technically and mathematically incorrect, it is a good way to think about it. The formula for standard deviation can be broken down into the following steps:

1. Find the mean of the data.
2. For each number in the dataset, subtract it from the mean and then square it.
3. Find the average of each square difference.
4. Take the square root of the number obtained in step three. This is the standard deviation.

Notice how, in the steps, we do actually take an arithmetic mean as one of the steps.

For example, look back at the temperature dataset. Let's find the standard deviation of the dataset using Python:

```
import numpy

temps = [31, 32, 32, 31, 28, 29, 31, 38, 32, 31, 30, 29, 30, 31, 26]

mean = numpy.mean(temps)     # == 30.73

squared_differences = []
# empty list o squared differences

for temperature in temps:
    difference = temperature - mean
 # how far is the point from the mean

    squared_difference = difference**2
    # square the difference

    squared_differences.append(squared_difference)
    # add it to our list


average_squared_difference = numpy.mean(squared_differences)
# This number is also called the "Variance"


standard_deviation = numpy.sqrt(average_squared_difference)
# We did it!


print standard_deviation  # == 2.5157
```

All of this code led to us find out that the standard deviation of the dataset is around 2.5, meaning that "on average", a data point is 2.5 degrees off from the average temperature of around 31 degrees, meaning that the temperature could likely dip below 29 degrees again in the near future.

The reason we want the "square difference" between each point and the mean and not the "actual difference" is because squaring the value actually puts emphasis on outliers—data points that are abnormally far away.

Measures of variation give us a very clear picture of how spread out or dispersed our data is. This is especially important when we are concerned with ranges of data and how data can fluctuate (think percent return on stocks).

The big difference between data at this level and at the next level lies in something that is not obvious.

Data at the interval level does not have a "natural starting point or a natural zero". However, being at zero degrees Celsius does not mean that you have "no temperature".

The ratio level

Finally, we will take a look at the ratio level. After moving through three different levels with differing levels of allowed mathematical operations, the ratio level proves to be the strongest of the four.

Not only can we define order and difference, the ratio level allows us to *multiply and divide* as well. This might seem like not much to make a fuss over but it changes almost everything about the way we view data at this level.

Examples

While Fahrenheit and Celsius are stuck in the interval level, the Kelvin scale of temperature boasts a natural zero. A measurement zero Kelvin literally means the absence of heat. It is a non-arbitrary starting zero. We can actually scientifically say that 200 Kelvin is twice as much heat as 100 Kelvin.

Money in the bank is at the ratio level. You can have "no money in the bank" and it makes sense that $200,000 is "twice as much as" $100,000.

Many people may argue that Celsius and Fahrenheit also have a starting point (mainly because we can convert from Kelvin to either of the two). The real difference here might seem silly, but because the conversion to Celsius and Fahrenheit make the calculations go into the negative, it does not define a clear and "natural" zero.

Measures of center

The arithmetic mean still holds meaning at this level, as does a new type of mean called the **geometric mean**. This measure is generally not used as much even at the ratio level, but is worth mentioning. It is the square root of the product of all the values.

For example, in our fridge temperature data, we can calculate the geometric mean as shown here:

```
import numpy

temps = [31, 32, 32, 31, 28, 29, 31, 38, 32, 31, 30, 29, 30, 31, 26]

num_items = len(temps)
product = 1.

for temperature in temps:
    product *= temperature

geometric_mean = product**(1./num_items)

print geometric_mean   # == 30.634
```

Note again how it is close to the arithmetic mean and median as calculated before. This is not always the case, and will be talked about at great length in the statistics chapter of this book.

Problems with the ratio level

Even with all of this added functionality at this level, we must generally also make a very large assumption that actually makes the ratio level a bit restrictive.

Data at the ratio level is usually non-negative.

For this reason alone, many data scientists prefer the interval level to the ratio level. The reason for this restrictive property is because if we allowed negative values, the ratio might not always make sense.

Consider that we allowed debt to occur in our money in the bank example. If we had a balance of $50,000, the following ratio would not really make sense at all:

$$\frac{\$50,000}{-\$50,000} = -1$$

Data is in the eye of the beholder

It is possible to impose structure on data. For example, while I said that you technically cannot use a mean for the one to five data at the ordinal scale, many statisticians would not have a problem using this number as a descriptor of the dataset.

The level at which you are interpreting data is a *huge* assumption that should be made at the beginning of any analysis. If you are looking at data that is generally thought of at the ordinal level and applying tools such as the arithmetic mean and standard deviation, this is something that data scientists must be aware of. This is mainly because if you continue to hold these assumptions as valid in your analysis, you may encounter problems. For example, if you also assume divisibility at the ordinal level by mistake, you are imposing structure where structure may not exist.

Summary

The type of data that you are working with is a very large piece of data science. It must precede most of your analysis because the type of data you have impacts the type of analysis that is even possible!

Whenever you are faced with a new dataset, the first three questions you should ask about it are the following:

- Is the data organized or unorganized?

  For example, does our data exist in a nice, clean row/column structure?

- Is each column quantitative or qualitative?

  For example, are the values numbers, strings, or do they represent quantities?

- At what level of data is each column?

  For example, are the values at the nominal, ordinal, interval, or ratio level?

The answers to these questions will not only impact your knowledge of the data at the end, but will also dictate the next steps of your analysis. They will dictate the types of graphs you are able to use and how you interpret them in your upcoming data models. Sometimes we will have to convert from one level to another in order to gain more perspective. In the coming chapters, we will take a much deeper look at how to deal with and explore data at different levels.

By the end of this book, we will be able to not only recognize data at different levels, but will also know how to deal with it at these levels.

3
The Five Steps of Data Science

We have spent extensive time looking at the preliminaries of data science, including outlining the types of data and how to approach datasets depending on their type. This chapter will focus mostly on the third step of exploration. We will use the Python packages `pandas` and `matplotlib` to explore different datasets.

Introduction to data science

Many people ask me the biggest difference between data science and data analytics. While one can argue that there is no difference between the two, many will argue that there are hundreds! I believe that regardless of how many differences there are between the two terms, the biggest is that *data science follows a structured, step-by-step process that, when followed, preserves the integrity of the results.*

Like any other scientific endeavor, this process must be adhered to, or else the analysis and the results are in danger of scrutiny. On a simpler level, following a strict process can make it much easier for amateur data scientists to obtain results faster than if they were exploring data with no clear vision.

While these steps are a guiding lesson for amateur analysts, they also provide the foundation for all data scientists, even those in the highest levels of business and academia. Every data scientist recognizes the value of these steps and follows them in some way or another.

Overview of the five steps

The five essential steps to perform data science are as follows:

1. Asking an interesting question
2. Obtaining the data
3. Exploring the data
4. Modeling the data
5. Communicating and visualizing the results

First, let's look at the five steps with reference to the big picture.

Ask an interesting question

This is probably my favorite step. As an entrepreneur, I ask myself (and others) interesting questions every day. I would treat this step as you would treat a brainstorming session. Start writing down questions regardless of whether or not you think the data to answer these questions even exists. The reason for this is twofold. First off, you don't want to start biasing yourself even before searching for data. Secondly, obtaining data might involve searching in both public and private locations and, therefore, might not be very straightforward. You might ask a question and immediately tell yourself "Oh, but I bet there's no data out there that can help me," and cross it off your list. Don't do that! Leave it on your list.

Obtain the data

Once you have selected the question you want to focus on, it is time to scour the world for the data that might be able to answer that question. As mentioned before, the data can come from a variety of sources; so, this step can be very creative!

Explore the data

Once we have the data, we use the lessons learned in *Chapter 2, Types of Data,* of this book and begin to break down the types of data that we are dealing with. This is a pivotal step in the process. Once this step is completed, the analyst generally has spent several hours learning about the domain, using code or other tools to manipulate and explore the data, and has a very good sense of what the data might be trying to tell them.

Model the data

This step involves the use of statistical and machine learning models. In this step, we are not only fitting and choosing models, we are implanting mathematical validation metrics in order to quantify the models and their effectiveness.

Communicate and visualize the results

This is arguably the most important step. While it might seem obvious and simple, the ability to conclude your results in a digestible format is much more difficult than it seems. We will look at different examples of cases when results were communicated poorly and when they were displayed very well.

In this book, we will focus mainly on steps 3, 4 and 5.

Why are we skipping steps 1 and 2 in this book?

While the first two steps are undoubtedly imperative to the process, they generally precede statistical and programmatic systems. Later in this book, we will touch upon the different ways to obtain data, however, for the purpose of focusing on the more scientific aspects of the process, we will begin with exploration right away.

Explore the data

The process of exploring data is not defined simply. It involves the ability to recognize the different types of data, transform data types, and use code to systemically improve the quality of the entire dataset to prepare it for the modeling stage. In order to best represent and teach the art of exploration, I will present several different datasets and use the python package pandas to explore the data. Along the way, we will run into different tips and tricks for how to handle data.

There are three basic questions we should ask ourselves when dealing with a new dataset that we may not have seen before. Keep in mind that these questions are not the beginning and the end of data science; they are some guidelines that should be followed when exploring a newly obtained set of data.

Basic questions for data exploration

When looking at a new dataset, whether it is familiar to you or not, it is important to use the following questions as guidelines for your preliminary analysis:

- Is the data organized or not?

  We are checking for whether or not the data is presented in a row/column structure. For the most part, data will be presented in an organized fashion. In this book, over 90% of our examples will begin with organized data. Nevertheless, this is the most basic question that we can answer before diving any deeper into our analysis.

  A general rule of thumb is that if we have unorganized data, we want to transform it into a row/column structure. For example, earlier in this book, we looked at ways to transform text into a row/column structure by counting the number of words/phrases.

- What does each row represent?

  Once we have an answer to how the data is organized and are now looking at a nice row/column based dataset, we should identify what each row actually represents. This step is usually very quick, and can help put things in perspective much more quickly.

- What does each column represent?

  We should identify each column by the level of data and whether or not it is quantitative/qualitative, and so on. This categorization might change as our analysis progresses, but it is important to begin this step as early as possible.

- Are there any missing data points?

  Data isn't perfect. Sometimes we might be missing data because of human or mechanical error. When this happens, we, as data scientists, must make decisions about how to deal with these discrepancies.

- Do we need to perform any transformations on the columns?

  Depending on what level/type of data each column is at, we might need to perform certain types of transformations. For example, generally speaking, for the sake of statistical modeling and machine learning, we would like each column to be numerical. Of course, we will use Python to make any and all transformations.

All the while, we are asking ourselves the overall question, *what can we infer from the preliminary inferential statistics?* We want to be able to understand our data a bit more than when we first found it.

Enough talk, let's see an example in the following section.

Dataset 1 – Yelp

The first dataset we will look at is a public dataset made available by the restaurant review site, Yelp. All personally identifiable information has been removed. Let's read in the data first, as shown here:

```
import pandas as pd

yelp_raw_data = pd.read_csv("yelp.csv")

yelp_raw_data.head()
```

A quick recap of what the preceding code does:

- Import the `pandas` package and nickname it as `pd`.
- Read in the `.csv` from the Web; call is `yelp_raw_data`.
- Look at the `head` of the data (just the first few rows).

	business_id	date	review_id	stars	text	type	user_id	cool	useful	funny
0	9yKzy9PApeiPPOUJEtnvkg	2011-01-26	fWKvX83p0-ka4JS3dc6E5A	5	My wife took me here on my birthday for breakf...	review	rLtl8ZkDX5vH5nAx9C3q5Q	2	5	0
1	ZRJwVLyzEJq1VAihDhYiow	2011-07-27	IjZ33sJrzXqU-0X6U8NwyA	5	I have no idea why some people give bad review...	review	0a2KyEL0d3Yb1V6aivbIuQ	0	0	0
2	6oRAC4uyJCsJl1X0WZpVSA	2012-06-14	IESLBzqUCLdSzSqm0eCSxQ	4	love the gyro plate. Rice is so good and I als...	review	0hT2KtfLiobPvh6cDC8JQg	0	1	0
3	_1QQZuf4zZOyFCvXc0o6Vg	2010-05-27	G-WvGaISbqqaMHlNnByodA	5	Rosie, Dakota, and I LOVE Chaparral Dog Park!!...	review	uZetl9T0NcROGOyFfughhg	1	2	0
4	6ozycU1RpktNG2-1BroVtw	2012-01-05	1uJFq2r5QfJG_6ExMRCaGw	5	General Manager Scott Petello is a good egg!!!...	review	vYmM4KTsC8ZfQBg-j5MWkw	0	0	0

Is the data organized or not?

- Because we have a nice row/column structure, we can conclude that this data seems pretty organized.

What does each row represent?

- It seems pretty obvious that each row represents a user giving a review of a business. The next thing we should do is to examine each row and label it by the type of data it contains. At this point, we can also use python to figure out just how big our dataset is. We can use the `shape` quality of a Dataframe to find this out, as shown:

```
yelp_raw_data.shape

# (10000,10)
```

- It tells us that this dataset has `10000` rows and `10` columns. Another way to say this is that this dataset has 10,000 observations and 10 characteristics.

What does each column represent?

Note that we have `10` columns:

- `business_id`: This is likely a unique identifier for the business the review is for. This would be at the **nominal level** because there is no natural order to this identifier.
- `date`: This is probably the date at which the review was posted. Note that it seems to be only specific to the day, month, and year. Even though time is usually considered continuous, this column would likely be considered discrete and at the **ordinal level** because of the natural order that dates have.
- `review_id`: This is likely a unique identifier for the review that each post represents. This would be at the **nominal level** because, again, there is no natural order to this identifier.
- `stars`: From a quick look (don't worry; we will perform some further analysis soon), we can see that this is an ordered column that represents what the reviewer gave the restaurant as a final score. This is ordered and qualitative; so, this is at the **ordinal level**.
- `text`: This is likely the raw text that each reviewer wrote. As with most text, we place this at the **nominal level**.
- `type`: In the first five columns, all we see is the word *review*. This might be a column that identifies that each row is a review, implying that there might be another type of row other than a review. We will take a look at this later. We place this at the **nominal level**.
- `user_id`: This is likely a unique identifier for the user who is writing the review. Just like the other unique IDs, we place this data at the **nominal level**.

Note that after we have looked at all of the columns, and found that all of the data is either at the ordinal level or at the nominal level, we have to look at the following things. This is not uncommon, but it is worth mentioning.

Are there any missing data points?

- Perform an isnull operation. For example if your dataframe is called awesome_dataframe then try the python command `awesome_dataframe.isnull().sum()` which will show the number of missing values in each column.

Do we need to perform any transformations on the columns?

- At this point, we are looking for a few things. For example, will we need to change the scale of some of the quantitative data, or do we need to create dummy variables for the qualitative variables? As this dataset has only qualitative columns, we can only focus on transformations at the ordinal and nominal scale.

Before starting, let's go over some quick terminology for pandas, the python data exploration module.

Dataframes

When we read in a dataset, Pandas creates a custom object called **Dataframe**. Think of this as the python version of a spreadsheet (but way better). In this case, the variable, `yelp_raw_data`, is a Dataframe.

To check whether this is true in Python, type in the following code:

```
type(yelp_raw_data)

# pandas.core.frame.DataFrame
```

Dataframes are two-dimensional in nature, meaning that they are organized in a row/column structure just as a spreadsheet is. The main benefits of using Dataframes over, say, a spreadsheet software would be that a Dataframe can handle much larger data than most common spreadsheet software. If you are familiar with the R language, you might recognize the word Dataframe. This is because the name was actually borrowed from the language!

As most of the data that we will deal with is organized, Dataframes are likely the most used object in pandas, second only to the Series object.

Series

The **Series** object is simply a Dataframe, but only with one dimension. Essentially, it is a list of data points. Each column of a Dataframe is considered to be a Series object. Let's check this. The first thing we need to do is grab a single column from our Dataframe; we generally use what is known as `bracket notation`. The following is an example:

```
yelp_raw_data['business_id'] # grabs a single column of the Dataframe
```

We will list the first few and last few rows:

```
0       9yKzy9PApeiPPOUJEtnvkg
1       ZRJwVLyzEJq1VAihDhYiow
2       6oRAC4uyJCsJl1X0WZpVSA
3       _1QQZuf4zZOyFCvXc0o6Vg
4       6ozycU1RpktNG2-1BroVtw
5       -yxfBYGB6SEqszmxJxd97A
6       zp713qNhx8d9KCJJnrw1xA
```

Let's use the type function to check that this column is a Series:

```
type(yelp_raw_data['business_id'])

# pandas.core.series.Series
```

Exploration tips for qualitative data

Using these two Pandas objects, let's start performing some preliminary data exploration. For qualitative data, we will specifically look at the nominal and ordinal levels.

Nominal level columns

As we are at the nominal level, let's recall that at this level, data is qualitative and is described purely by name. In this dataset, this refers to the `business_id`, `review_id`, `text`, `type`, and `user_id`. Let's use Pandas in order to dive a bit deeper, as shown here:

```
yelp_raw_data['business_id'].describe()

# count                       10000
# unique                       4174
# top       JokKtdXU7zXHcr20Lrk29A
# freq                           37
```

The `describe` function will give us some quick stats about the column whose name we enter into the quotation marks. Note how Pandas automatically recognized that `business_id` was a qualitative column and gave us stats that make sense. When `describe` is called on a qualitative column, we will always get the following four items:

- `count`: How many values are filled in
- `unique`: How many unique values are filled in
- `top`: The name of the most common item in the dataset
- `freq`: How often the most common item appears in the dataset

At the nominal level, we are usually looking for a few things, that would signal a transformation:

- Do we have a reasonable number (usually under 20) of unique items?
- Is this column free text?
- Is this column completely unique across all rows?

So, for the `business_id` column, we have a count of `10000`. Don't be fooled though! This does not mean that we have 10,000 businesses being reviewed here. It just means that of the 10,000 rows of reviews, the `business_id` column is filled in all 10,000 times. The next qualifier, `unique`, tells us that we have `4174` unique businesses being reviewed in this dataset. The most reviewed business is business `JokKtdXU7zXHcr20Lrk29A`, which was reviewed `37` times.

```
yelp_raw_data['review_id'].describe()

# count                    10000
# unique                   10000
# top       eTa5KD-LTgQv6UT1Zmijmw
# freq                         1
```

We have a `count` of `10000` and a `unique` of `10000`. Think for a second, does this make sense? Think about what each row represents and what this column represents.

(insert jeopardy theme song here)

Of course it does! Each row of this dataset is supposed to represent a single, unique review of a business and this column is meant to serve as a unique identifier for a review; so, it makes sense that the `review_id` column has `10000` unique items in it. So, why is `eTa5KD-LTgQv6UT1Zmijmw` the *most common* review? This is just a random choice from the 10,000 and means nothing.

```
yelp_raw_data['text'].describe()

count                                                   10000
unique                                                   9998
top       This review is for the chain in general. The l...
freq                                                        2
```

This column, which represents the actual text people wrote, is interesting. We would imagine that this should also be similar to `review_id` in that there should be all unique text, because it would be weird if two people wrote exactly the same thing; but we have two reviews with the exact same text! Let's take a second to learn about Dataframe filtering to examine this further.

Filtering in Pandas

Let's talk a bit about how filtering works. Filtering rows based on certain criteria is quite easy in Pandas. In a Dataframe, if we wish to filter out rows based on some search criteria, we will need to go row by row and check whether or not a row satisfies that particular condition. Pandas handles this by passing in a Series of *Trues* and *Falses* (Booleans).

We literally pass into the Dataframe a list of `True` and `False` data that mean the following:

- `True`: This row satisfies the condition
- `False`: This row does not satisfy the condition

So, first let's make the conditions. In the following lines of code, I will grab the text that occurs twice:

```
duplicate_text = yelp_raw_data['text'].describe()['top']
```

Here is a snippet of the text:

```
"This review is for the chain in general. The location we went to is
new so it isn't in Yelp yet. Once it is I will put this review there
as well……."
```

Right off the bat, we can guess that this might actually be one person who went to review two businesses that belong to the same chain and wrote the exact same review. However, this is just a guess right now.

The `duplicate_text` variable is of `string` type.

Now that we have this text, let's use some magic to create that Series of true and false:

```
text_is_the_duplicate = yelp_raw_data['text'] == duplicate_text
```

Right away you might be confused. What we have done here is take the text column of the Dataframe and compared it to the string, `duplicate_text`. This is strange because we seem to be comparing a list of 10,000 elements to a single string. Of course, the answer should be a straight false, right?

The Pandas' Series has a very interesting feature in that if you compare the Series to an object, it will return another Series of Booleans of the same length where each true and false is the answer to the question, *is this element the same as the element you are comparing it to?* Very handy!

```
type(text_is_the_duplicate) # it is a Series of Trues and Falses

text_is_the_duplicate.head() # shows a few Falses out of the Series
```

In Python, we can add and subtract true and false as if they were 1 and 0, respectively. For example, *True + False – True + False + True == 1*. So, we can verify that this Series is correct by adding up all of the values. As only two of these rows should contain the duplicate text, the sum of the Series should only be `2`, which it is! This is shown as follows:

```
sum(text_is_the_duplicate) # == 2
```

Now that we have our Series of Booleans, we can pass it directly into our Dataframe, using bracket notation, and get our filtered rows, as illustrated:

```
filtered_dataframe = yelp_raw_data[text_is_the_duplicate]
# the filtered Dataframe

filtered_dataframe
```

	business_id	date	review_id	stars	text	type	user_id	cool	useful	funny
4372	jvvh4Q00Hq2XylcfmAAT2A	2012-06-16	ivGRamFF3KurE9bjkl6uMw	2	This review is for the chain in general. The l...	review	KLekdmo4FdNnP0huUhzZNw	0	0	0
9680	rlonUa02zMz_ki8eF-Adug	2012-06-16	mutQE6UfjLlpJ8Wozpq5UA	2	This review is for the chain in general. The l...	review	KLekdmo4FdNnP0huUhzZNw	0	0	0

It seems that our suspicions were correct and one person, on the same day, gave the exact same review to two different `business_id`, presumably a part of the same chain. Let's keep moving along to the rest of our columns:

```
yelp_raw_data['type'].describe()


count       10000
unique          1
top        review
freq        10000
```

Remember this column? Turns out they are all the exact same type, namely `review`.

```
yelp_raw_data['user_id'].describe()
count                        10000
unique                        6403
top         fczQCSmaWF78toLEmb0Zsw
freq                            38
```

Similar to the `business_id` column, all the `10000` values are filled in with `6403` unique users and one user reviewing `38` times!

In this example, we won't have to perform any transformations.

Ordinal level columns

As far as ordinal columns go, we are looking at date and `stars`. For each of these columns, let's look at what the describe method brings back:

```
yelp_raw_data['stars'].describe()
# count    10000.000000
# mean         3.777500
# std          1.214636
# min          1.000000
# 25%          3.000000
# 50%          4.000000
# 75%          5.000000
# max          5.000000
```

Woah! Even though this column is ordinal, the describe method returned stats that we might expect for a quantitative column. This is because the software saw a bunch of numbers and just assumed that we wanted stats like the mean or the min and max. This is no problem. Let's use a method called value_counts to see the count distribution, as shown here:

```
yelp_raw_data['stars'].value_counts()
# 4    3526
# 5    3337
# 3    1461
# 2     927
# 1     749
```

The value_counts method will return the distribution of values for any column. In this case, we see that the star rating 4 is the most common, with 3526 values, followed closely by the rating 5. We can also plot this data to get a nice visual. First, let's sort by star rating, and then use the prebuilt plot method to make a bar chart.

```
dates = yelp_raw_data['stars'].value_counts()
dates.sort()
dates.plot(kind='bar')
```

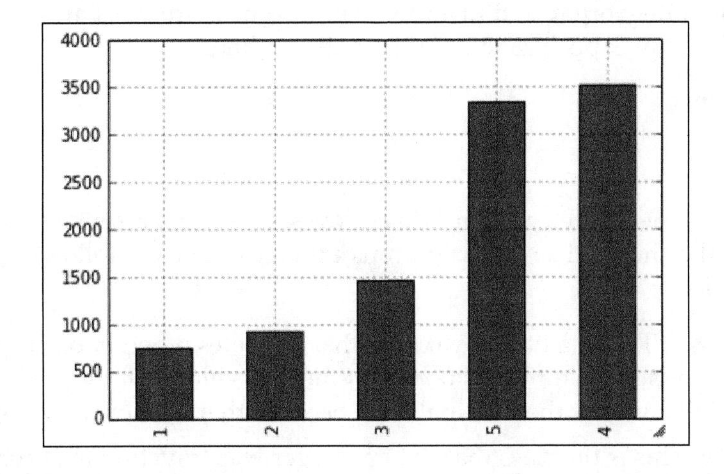

People are definitely more likely to give good star ratings over bad ones! We can follow this procedure for the date column. I will leave you to try it on your own. For now, let's look at a new dataset.

Dataset 2 – titanic

The `titanic` dataset contains a sample of people who were on the Titanic when it struck an iceberg in 1912. Let's go ahead and import it, as shown here:

```
titanic = pd.read_csv('short_titanic.csv')
titanic.head()
```

	Survived	Pclass	Name	Sex	Age
0	0	3	Braund, Mr. Owen Harris	male	22
1	1	1	Cumings, Mrs. John Bradley (Florence Briggs Th...	female	38
2	1	3	Heikkinen, Miss. Laina	female	26
3	1	1	Futrelle, Mrs. Jacques Heath (Lily May Peel)	female	35
4	0	3	Allen, Mr. William Henry	male	35

This Dataframe usually has more columns; however, for our example, we will only focus on the given columns.

This data is definitely organized in a row/column structure, as is most spreadsheet data. Let's take a quick peek as to its size, as shown here:

```
titanic.shape

# (891, 5)
```

So, we have `891` rows and `5` columns. Each row seems to represent a single passenger on the ship and as far as columns are concerned, the following list tells us what they indicate:

- `Survived`: This is a binary variable that indicates whether or not the passenger survived the accident (`1` if they survived, `0` if they died). This would likely be at the nominal level because there are only two options.
- `Pclass`: This is the class that the passenger was traveling in (`3` for third class, and so on). This is at the ordinal level.
- `Name`: This is the name of the passenger, and it is definitely at the nominal level.

- `Sex`: This indicates the gender of the passenger. It is at the nominal level.
- `Age`: This one is a bit tricky. Arguably, you may place age at either a qualitative or quantitative level, however, I think that age belongs to a quantitative state, and thus, to the ratio level.

As far as transformations are concerned, usually, we want all columns to be numerical, regardless of their qualitative state. This means that `Name` and `Sex` will have to be converted into numerical columns somehow. For `Sex`, we can change the column to hold `1` if the passenger was female and `0` if they were male. Let's use Pandas to make the change. We will have to import another Python module, called `numpy` or numerical Python, as illustrated:

```
import numpy as np
titanic['Sex'] = np.where(titanic['Sex']=='female', 1, 0)
```

The `np.where` method takes in three things:

- A list of Booleans (true or false)
- A new value
- A backup value

The method will replace all true with the first value (in this case `1`) and the false with the second value (in this case `0`), leaving us with a new numerical column that represents the same thing as the original `Sex` column.

```
titanic['Sex']

# 0      0
# 1      1
# 2      1
# 3      1
# 4      0
# 5      0
# 6      0
# 7      0
```

Let's use a shortcut and describe all the columns at once, as shown:

```
titanic.describe()
```

	Survived	Pclass	Sex	Age
count	891.000000	891.000000	891.000000	714.000000
mean	0.383838	2.308642	0.352413	29.699118
std	0.486592	0.836071	0.477990	14.526497
min	0.000000	1.000000	0.000000	0.420000
25%	0.000000	2.000000	0.000000	20.125000
50%	0.000000	3.000000	0.000000	28.000000
75%	1.000000	3.000000	1.000000	38.000000
max	1.000000	3.000000	1.000000	80.000000

Note how our qualitative columns are being treated as quantitative; however, I'm looking for something irrelevant to the data type. Note the count row: `Survived`, `Pclass`, and `Sex` all have `891` values (the number of rows), but `Age` only has `714` values. Some are missing! To double verify, let's use the Pandas functions, called `isnull` and `sum`, as shown:

```
titanic.isnull().sum()

Survived        0
Pclass          0
Name            0
Sex             0
Age           177
```

This will show us the number of missing values in each column. So, `Age` is the only column with missing values to deal with.

When dealing with missing values, you usually have the following two options:

- Drop the row with the missing value
- Try to fill it in

Dropping the row is the easy choice; however, you run the risk of losing valuable data! For example, in this case, we have `177` missing age values (891-714) which is nearly 20% of the data. To fill in the data, we could either go back to the history books, find each person one by one, and fill in their age, or we can fill in the age with a placeholder value.

Let's fill in each missing value of the `Age` column with the overall average age of the people in the dataset. For this, we will use two new methods, called `mean` and `fillna`. We use `isnull` to tell us which values are null and the `mean` function to give us the average value of the `Age` column. `fillna` is a Pandas method that replaces null values with a given value.

```
print sum(titanic['Age'].isnull()) # == 177 missing values

average_age = titanic['Age'].mean() # get the average age

titanic['Age'].fillna(average_age, inplace = True) #use the fillna
method to remove null values

print sum(titanic['Age'].isnull()) # == 0 missing values
```

We're done! We have replaced each value with `26.69`, the average age in the dataset.

```
titanic.isnull().sum()

Survived      0
Pclass        0
Name          0
Sex           0
Age           0
```

Great! Nothing is missing, and we did not have to remove any rows.

```
titanic.head()
```

	Survived	Pclass	Name	Sex	Age
0	0	3	Braund, Mr. Owen Harris	0	22
1	1	1	Cumings, Mrs. John Bradley (Florence Briggs Th...	1	38
2	1	3	Heikkinen, Miss. Laina	1	26
3	1	1	Futrelle, Mrs. Jacques Heath (Lily May Peel)	1	35
4	0	3	Allen, Mr. William Henry	0	35

At this point, we could start getting a bit more complicated with our questions. For example, *what is the average age for a female or a male?* To answer this, we can filter by each gender and take the mean age. Pandas has a built-in function for this, called `groupby`, as illustrated here:

```
titanic.groupby('Sex')['Age'].mean()
```

This means *group the data by the Sex column, and then give me the mean of age for each group*. This gives us the following output:

```
Sex
0       30.505824
1       28.216730
```

We will ask more of these difficult and complex questions and will be able to answer them with Python and statistics.

Summary

Though this is only our first look at data exploration, don't worry—this is definitely not the last time we will follow these steps for data science and exploration. From now on, every time we look at a new piece of data, we will use our steps of exploration to transform, break down, and standardize our data. The steps outlined in this chapter, while they are only guidelines, form a standard practice that any data scientist can follow in their work. The steps can also be applied to any dataset that requires analysis.

We are rapidly approaching the section of the book that deals with statistical, probabilistic, and machine learning models. Before we can truly jump into these models, we have to look at some of the basics of mathematics. In the next chapter, we will take a look at some of the math necessary to perform some of the more complicated operations in modeling, but don't worry—the math required for this process is minimal, and we will go through it step by step.

4
Basic Mathematics

It's time to start looking at some basic mathematic principles that are handy when dealing with data science. The word *math* tends to strike fear in the hearts of many, but I aim to make this as enjoyable as possible. In this chapter, we will go over the basics of the following topics:

- Basic symbols/terminology
- Logarithms/exponents
- The set theory
- Calculus
- Matrix (linear) algebra

We will also cover other fields of mathematics. Moreover, we will see how to apply each of these to various aspects of data science as well as other scientific endeavors.

Recall that, in a previous chapter, we identified math as being one of the three key components of data science. In this chapter, I will introduce concepts that will become important later on in this book—when looking at probabilistic and statistical models—and I will also be looking at concepts that will be useful in this chapter. Regardless of this, all of the concepts in this chapter should be considered fundamentals in your quest to become a data scientist.

Mathematics as a discipline

Mathematics, as a science, is one of the oldest known forms of logical thinking by mankind. Since ancient Mesopotamia and likely before (3,000 BCE), humans have been relying on arithmetic and more challenging forms of math to answer life's biggest questions.

Today, we rely on math for most aspects of our daily life; yes, I know that sounds cliché, but I mean it. Whether you are watering your plants or feeding your dog, your internal mathematical engine is constantly spinning—calculating how much water the plant had per day over the last week and predicting the next time your dog will be hungry given that they eat right now. Whether or not you are consciously using the principles of math, the concepts live deep inside everyone's brains. It's my job as a math teacher to get you to realize it.

Basic symbols and terminology

First, let's take a look at the most basic symbols that are used in the mathematical process as well as some more subtle notations used by data scientists.

Vectors and matrices

A **vector** is defined as an object with both magnitude and direction. This definition, however, is a bit complicated for our use. For our purpose, a vector is simply a 1-dimensional array representing a series of numbers. Put in another way, a vector is a list of numbers.

It is generally represented using an arrow or bold font, as shown:

$$\vec{x} \quad or \quad x$$

Vectors are broken into components, which are individual members of the vector. We use index notations to denote the element that we are referring to, as illustrated:

If $\vec{x} = \begin{pmatrix} 3 \\ 6 \\ 8 \end{pmatrix}$ then $x_1 = 3$

In math, we generally refer to the first element as index `1`, as opposed to computer science, where we generally refer to the first element as index `0`. It is important to remember what index system you are using.

In Python, we can represent arrays in many ways. We could simply use a Python list to represent the preceding array:

```
x = [3, 6, 8]
```

However, it is better to use the numpy array type to represent arrays, as shown, because it gives us much more utility when performing vector operations:

```
import numpy as np
x = np.array([3, 6, 8])
```

Regardless of the Python representation, vectors give us a simple way of storing *multiple dimensions* of a single data point/observation.

Consider that we measure the average satisfaction rating (0-100) of employees for three departments of a company as being *57* for HR, *89* for engineering, and *94* for management. We can represent this as a vector with the following formula:

$$x = \begin{pmatrix} x_1 \\ x_2 \\ x_3 \end{pmatrix} = \begin{pmatrix} 57 \\ 89 \\ 94 \end{pmatrix}$$

This vector holds three different bits of information about our data. This is the perfect use of a vector in data science.

You can also think of a vector as being the theoretical generalization of Panda's Series object. So, naturally, we need something to represent the Dataframe.

We can extend our notion of an array to move beyond a single dimension and represent data in multiple dimensions.

A **matrix** is a 2-dimensional representation of arrays of numbers. Matrices (plural) have two main characteristics that we need to be aware of. The dimension of the matrix, denoted as *n* x *m* (*n by m*), tells us that the matrix has *n* rows and *m* columns. Matrices are generally denoted using a capital, bold-faced letter, such as X. Consider the following example:

$$\begin{pmatrix} 3 & 4 \\ 8 & 55 \\ 5 & 9 \end{pmatrix}$$

This is a *3* x 2 (*3 by* 2) matrix because it has three rows and two columns.

If a matrix has the same number of rows and columns, it is called a **square matrix**.

The matrix is our generalization of the Pandas Dataframe. It is arguably one of the most important mathematical objects in our toolkit. It is used to hold organized information, in our case, data.

Revisiting our previous example, let's say we have three offices in different locations, each with the same three departments: HR, engineering, and management. We could make three different vectors, each holding a different office's satisfaction scores, as shown:

$$x = \begin{pmatrix} 57 \\ 89 \\ 94 \end{pmatrix}, y = \begin{pmatrix} 67 \\ 87 \\ 84 \end{pmatrix}, z = \begin{pmatrix} 65 \\ 98 \\ 60 \end{pmatrix}$$

However, this is not only cumbersome, but also unscalable. What if you have 100 different offices? Then we would need to have 100 different 1-dimensional arrays to hold this information.

This is where a matrix alleviates this problem. Let's make a matrix where each row represents a different department and each column represents a different office, as shown:

	Office 1	Office 2	Office 3
HR	57	67	65
Engineering	89	87	98
Management	94	84	60

This is much more natural. Now, let's strip away the labels, and we are left with a matrix!

$$X = \begin{pmatrix} 57 & 67 & 65 \\ 89 & 87 & 98 \\ 94 & 84 & 60 \end{pmatrix}$$

Quick exercises

1. If we add a fourth office, would we need a new row or column?
2. What would the dimension of the matrix be after we add the fourth office?

3. If we eliminate the management department from the original X matrix, what would the dimension of the new matrix be?
4. What is the general formula to know the number of elements in the matrix?

Answers

1. Column.
2. 3×4.
3. 2×3.
4. $n \times m$ (n being the number of rows and m being the number of columns).

Arithmetic symbols

In this section, we will go over some symbols associated with basic arithmetic that appear in most, if not all, data science tutorials and books.

Summation

The uppercase sigma Σ symbol is a universal symbol for addition. Whatever is to the right of the sigma symbol is usually something iterable, meaning that we can go over it one by one (for example, a vector).

For example, let's create the representation of a vector:

```
X = [1, 2, 3, 4, 5]
```

To find the sum of the content, we can use the following formula:

$$\sum x_i = 15$$

In Python, we can use the following formula:

```
sum(x) # == 15
```

For example, the formula for calculating the mean of a series of numbers is quite common. If we have a vector (x) of length n, the mean of the vector can be calculated as follows:

$$mean = \frac{1}{n} \sum x_i$$

This means that we will add up each element of *x*, denoted by x_i, and then multiply the sum by *1/n*, otherwise known as dividing by *n*, the length of the vector.

Proportional

The lowercase alpha symbol α represents values that are proportional to each other. This means that as one value changes, so does the other. The direction in which the values move depends on how the values are proportional. Values can either vary directly or indirectly. If values vary directly, they both move in the same direction (as one goes up, so does the other). If they vary indirectly, they move in opposite directions (if one goes down, the other goes up).

Consider the following examples:

- The sales of a company vary directly with the number of customers. This can be written as $Sales \, \alpha \, Customers$.
- Gas prices vary (usually) indirectly with oil availability, meaning that as the availability of oil goes down (it's more scarce), gas prices will go up. This can be denoted as $Gas \, \alpha \, Oil \, Availability$.

Later on, we will see a very important formula called the **Bayes formula**, which includes a variation symbol.

Dot product

The dot product is an operator like addition and multiplication. It is used to combine two vectors, as shown:

$$\begin{pmatrix} 3 \\ 7 \end{pmatrix} \cdot \begin{pmatrix} 9 \\ 5 \end{pmatrix} = 3*9+7*5=62$$

So, what does this mean? Let's say we have a vector that represents a customer's sentiments towards three genres of movies—comedy, romantic, and action.

When using a dot product, note that the answer is a single number, known as a **scalar**.

Consider that, on a scale of 1-5, a customer loves comedies, hates romantic movies, and is alright with action movies. We might represent this as follows:

$$\begin{pmatrix} 5 \\ 1 \\ 3 \end{pmatrix}$$

Here:

- *5* denotes the love for comedies,
- *1* is the hatred for romantic
- *3* is the indifference of action

Now, let's assume that we have two new movies, one of which is a romantic comedy and the other is a funny action movie. The movies would have their own vector of qualities, as shown:

$$m_1 = \begin{pmatrix} 4 \\ 5 \\ 1 \end{pmatrix} \text{ and } m_2 = \begin{pmatrix} 5 \\ 1 \\ 5 \end{pmatrix}$$

Here, m_1 is our romantic comedy and m_2 is our funny action movie.

In order to make a recommendation, we will *apply* the dot product between the customer's preferences for each movie. The higher value will win and, therefore, will be recommended to the user.

Let's compute the recommendation score for each movie. For movie *1*, we want to compute:

$$\begin{pmatrix} 5 \\ 1 \\ 3 \end{pmatrix} . \begin{pmatrix} 4 \\ 5 \\ 1 \end{pmatrix}$$

We can think of this problem as such:

Customer: M_1

$$\begin{pmatrix}5\\1\\3\end{pmatrix} \cdot \begin{pmatrix}4\\5\\1\end{pmatrix} = \begin{matrix}(5.4) \rightarrow \text{user loves comedies and this move is funny}\\ + \\ (1.5) \rightarrow \text{user hates romance but this move is romantic} \\ + \\ (3.1) \rightarrow \text{user doesn't mind action and the move is not action packed} \\ \hline 28\end{matrix}$$

The answer we obtain is *28*, but what does this number mean? On what scale is it? Well, the best score anyone can ever get is when all values are *5*, making the outcome as follows:

$$\begin{pmatrix}5\\5\\5\end{pmatrix} . \begin{pmatrix}5\\5\\5\end{pmatrix} = 5^2 + 5^2 + 5^2 = 75$$

The lowest possible score is when all values are *1*, as shown:

$$\begin{pmatrix}1\\1\\1\end{pmatrix} . \begin{pmatrix}1\\1\\1\end{pmatrix} = 1^2 + 1^2 + 1^2 = 3$$

So, we must think about *28* on a scale from *3-75*. To do this, imagine a number line from *3* to *75* and where *28* would be on it. This is illustrated as follows:

3 28 75

Not that far. Let's try for movie *2*:

$$\begin{pmatrix}5\\1\\3\end{pmatrix} . \begin{pmatrix}5\\1\\5\end{pmatrix} = (5*5)+(1*1)+(3*5) = 41$$

This is higher than *28*! Putting this number on the same timeline as before, we can also visually observe that it is a much better score, as shown:

So, between movie *1* and movie 2, we will definitely recommend movie 2 to our user. This is, in essence, how most movie prediction engines work. They build a customer profile, which is represented as a vector. They then take a vector representation of each movie they have to offer, combine them with the customer profile (perhaps with a dot product), and make recommendations from there. Of course, most companies must do this on a much larger scale, which is where a particular field of mathematics, called **linear algebra**, can be very useful; we will look at it later in this chapter.

Graphs

No doubt you have encountered dozens, if not hundreds, of graphs in your life so far. I'd like to mostly talk about conventions with regard to graphs and notations.

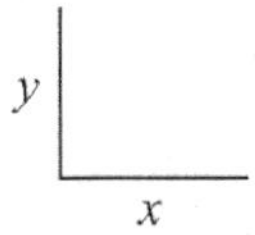

This is a basic **Cartesian graph** (x and y coordinate). The x and y notation are very standard but sometimes do not entirely explain the big picture. We sometimes refer to the x variable as being the independent variable and the y as the dependent variable. This is because when we write functions, we tend to speak about them as being *y is a function of x,* meaning that the value of y is dependent on the value of x. This is what a graph is trying to show.

Suppose we have two points on a graph, as shown:

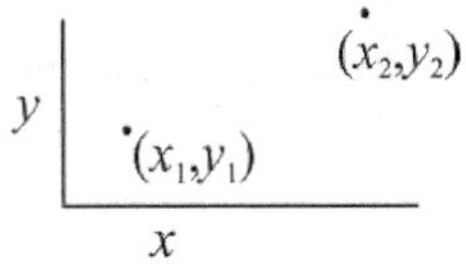

We refer to the points as (x_1, y_1) and (x_2, y_2).

The **slope** between these two points is defined as follows:

$$slope = m = \frac{y_2 - y_1}{x_2 - x_1}$$

You have probably seen this formula before, but it is worth mentioning if not for its significance. The slope defines the rate of change between the two points. Rates of change can be very important in data science, specifically in areas involving differential equations and calculus.

Rates of change are a way of representing how variables move together and to what degree. Consider that we are modeling the temperature of your coffee in relation to the time that it has been sitting out. Perhaps we have a rate of change as follows:

$$-\frac{2\,degrees\ F}{1\,minute}$$

This rate of change is telling us that for every single minute, our coffee's temperature is dropping by two degrees Fahrenheit.

Later on in this book, we will visit a machine learning algorithm, called linear regression. In linear regression, we are concerned with the rates of change between variables as they allow us to exploit this relationship for predictive purposes.

Think of the Cartesian plane as being an infinite plane of vectors with two elements. When people refer to higher dimensions, such as 3D or 4D, they are merely referring to an infinite space that holds vectors with more elements. A 3D space holds vectors of length three while a 7D space holds vectors with seven elements in them.

Logarithms/exponents

An **exponent** tells you how many times you have to multiply a number to itself, as illustrated:

exponent

$$2^4 = 2 \cdot 2 \cdot 2 \cdot 2 = 16$$

base

A **logarithm** is the number that answers the question: "what exponent gets me from the base to this other number?" This can be denoted as follows:

$$\log_2(16) = 4$$

base   logarithm

If these two concepts seem similar, then you are correct! Exponents and logarithms are heavily related. In fact, the words exponent and logarithm actually mean the same thing! A logarithm is an exponent. The preceding two equations are actually two versions of the same thing. The basic idea is that 2 times 2 times 2 times 2 is 16.

The following is a depiction of how we can use both versions to say the same thing. Note how I use arrows to move from the log formula to the exponent formula:

$$\log_2(16) = 4 \leftrightarrow 2^4 = 16$$

Consider the following examples:

- $\log_3 81 = 4$ because $3^4 = 81$
- $\log_5 125 = 3$ because $5^3 = 125$

Note something interesting, if we rewrite the first equation to be:

$$\log_3 81 = 4$$

We then replace 81 with the equivalent statement, 34, as follows:

$$\log_3 3^4 = 4$$

We can note something interesting: the 3s seem to *cancel out*. This is actually very important when dealing with numbers more difficult to work with than 3s and 4s.

Exponents and logarithms are most important when dealing with growth. More often than not, if some quantity is growing (or declining in growth), an exponent/logarithm can help model this behavior.

For example, the number *e* is around 2.718 and has many practical applications. A very common application is growth calculation for money. Suppose you have $5,000 deposited in a bank with continuously compounded interest at the rate of 3%, then we can use the following formula to model the growth of our deposit:

$$A = Pe^{rt}$$

Where:

- *A* denotes the final amount
- *P* denotes the principal investment (*5000*)
- *e* denotes constant (*2.718*)
- *r* denotes the rate of growth (*.03*)
- *t* denotes the time (in years)

We are curious, when will our investment double? How long would I have to have my money in this investment to achieve 100% growth? Basically:

$$10000 = 5000e^{.03t}$$

Is the formula we wish to solve:

$$10000 = 5000e^{.03t}$$

$$2 = e^{.03t} \quad (divide\, by\, 5000\, on\, both\, sides)$$

At this point, we have a variable in the exponent that we want to solve. When this happens, we can use the logarithm notation to figure it out!

$$2 = e^{.03t} \leftrightarrow \log_e(2) = .03t$$

This leaves us with $\log_e(2) = .03t$.

When we are taking the logarithm of a number with a base of *e*, it is called a *natural logarithm*. We rewrite the logarithm to be as follows:

$$In(2) = .03t$$

Using a calculator (or Python), we find that *ln(2) = 0.69.*

$$0.69 = .03t$$

$$t = 2.31$$

This means that it would take *2.31* years to double our money.

Set theory

The set theory involves mathematical operations at a set level. It is sometimes thought of as a basic fundamental group of theorems that governs the rest of mathematics. For our purpose, we use the set theory in order to manipulate groups of elements.

A **set** is a collection of distinct objects.

That's it! A set can be thought of as a list in Python, but with no repeat objects. In fact, there even exists a set of objects in Python:

```
s = set()

s = set([1, 2, 2, 3, 2, 1, 2, 2, 3, 2])
# will remove duplicates from a list

s == {1, 2, 3}
```

> Note that, in Python, the curly braces—{, }—can denote either a set or a dictionary.
>
> Remember that a dictionary in Python is a set of key-value pairs, for example:

```
dict = {"dog": "human's best friend", "cat": "destroyer of world"}
dict["dog"]# == "human's best friend"
len(dict["cat"]) # == 18

# but if we try to create a pair with the same key as an existing key
```

```
dict["dog"] = "Arf"

dict
{"dog": "Arf", "cat": "destroyer of world"}
# It will override the previous value
# dictionaries cannot have two values for one key.
```

They share this notation because they share a quality in that sets cannot have duplicate elements, just as dictionaries cannot have duplicate keys.

The **magnitude** of a set is the number of elements in the set and is represented as follows:

$$|A| = magnitude\ of\ A$$

```
s  # == {1,2,3}
len(s) == 3 # magnitude of s
```

The concept of an empty set exists and is denoted by the character ϕ. This null set is said to have a magnitude of 0.

If we wish to denote that an element is within a set, we use the epsilon notation, as shown:

$$2 \in \{1,2,3\}$$

This means that the element 2 exists in the set of *1*, *2*, and *3*. If one set is entirely inside another set, we say that it is a **subset** of its larger counterpart.

$$A = \{1,5,6\}, B = \{1,5,6,7,8\}$$

$$A \subseteq B$$

So, *A* is a subset of *B* and *B* is called the superset of *A*. If *A* is a subset of *B* but *A* does not equal *B* (meaning that there is at least one element in *B* that is not in *A*), then *A* is called a proper subset of *B*.

Consider the following examples:

- A set of even numbers is a subset of all integers
- Every set is a subset, but not a proper subset, of itself

- A set of all tweets is a superset of English tweets

In data science, we use sets (and lists) to represent a list of objects and, often, to generalize the behavior of consumers. It is common to reduce a customer to a set of characteristics.

Consider that we are a marketing firm trying to predict where a person wants to shop for clothes. We are given a set of clothing brands the user has previously visited, and our goal is to predict a new store that they would also enjoy. Suppose a specific user has previously shopped at the following stores:

```
user1 = {"Target","Banana Republic","Old Navy"}
# note that we use {} notation to create a set
# compare that to using [] to make a list
```

So, `user1` has previously shopped at `Target`, `Banana Republic`, and `Old Navy`. Let's also look at a different user, called `user2`, as shown:

```
user2 = {"Banana Republic","Gap","Kohl's"}
```

Suppose we are wondering how similar these users are. With the limited information we have, one way to define similarity is to see how many stores there are that they both shop at. This is called an intersection.

The **intersection** of two sets is a set whose elements appear in both the sets. It is denoted using the symbol $\cap$, as shown:

$$user1 \cap user2 = \{Banana\ Republic\}$$

$$|user1 \cap user2| = 1$$

The intersection of the two users is just one store. So, right away that doesn't seem great. However, each user only has three elements in their set, so having 1/3 does not seem as bad. Suppose we are curious about how many stores are represented between the two of them; this is called a union.

The **union** of two sets is a set whose elements appear in either set. It is denoted using the symbol $\cup$, as shown:

$$user1 \cup user2 = \{Banana\ Republic, Target, Old\ Navy, Gap, Kohl's\}$$

$$|user1 \cup user2| = 5$$

When looking at the similarity between *user1* and *user2*, we should use a combination of the union and the intersection of their sets. *user1* and *user2* have one element in common out of a total of five distinct elements between them. So, we can define the similarity between the two users as follows:

$$\frac{|user1 \cap user2|}{|user1 \cup user2|} = \frac{1}{5} = .2$$

In fact, this has a name in the set theory. It is called the **jaccard measure**. In general, for the sets *A* and *B*, the jaccard measure (jaccard similarity) between the two sets is defined as follows:

$$JS(A,B) = \frac{|A \cap B|}{|A \cup B|}$$

It can also be defined as the magnitude of the intersection of the two sets divided by the magnitude of the union of the two sets.

This gives us a way to quantify similarities between elements represented with sets.

Intuitively, the jaccard measure is a number between 0 and 1, such that when the number is closer to 0, the people are more dissimilar and when the measure is closer to 1, the people are considered similar to each other.

If we think about the definition, then it actually makes sense. Take a look at the measure once more:

$$JS(A,B) = \frac{Number\ of\ stores\ they\ share\ in\ common}{Unique\ number\ of\ store\ they\ like\ combined}$$

Here, the numerator represents the number of stores that the users have in common (in the sense that they like shopping there), while the denominator represents the unique number of stores that they like put together.

We can represent this in Python using some simple code, as shown:

```
user1 = {"Target","Banana Republic","Old Navy"}
user2 = {"Banana Republic","Gap","Kohl's"}

def jaccard(user1, user2):
  stores_in_common = len(user1 & user2)
```

```
    stores_all_together = len(user1 | user2)
    return stores / float(stores_all_together)

# I cast stores_all_together as a float to return a decimal answer
instead of python's default integer division

# so
jaccard(user1, user2) == # 0.2 or 1/5
```

The set theory becomes highly prevalent when we enter the world of probability and also when dealing with high-dimensional data. We will use sets to represent real-world events taking place and probability becomes set theory with vocabulary on top of it.

Linear algebra

Remember the movie recommendation engine we looked at earlier? What if we had 10,000 movies to recommend and we had to choose only 10 to give to the user? We'd have to take a dot product between the user profile and each of the 10,000 movies. Linear algebra provides the tools to make these calculations much more efficient.

It is an area of mathematics that deals with the math of matrices and vectors. It has the aim of breaking down these objects and reconstructing them in order to provide practical applications. Let's look at a few linear algebra rules before proceeding.

Matrix multiplication

Like numbers, we can multiple matrices together. Multiplying matrices is, in essence, a mass produced way of taking several dot products at once. Let's, for example, try to multiple the following matrices:

$$\begin{pmatrix} 1 & 5 \\ 5 & 8 \\ 7 & 8 \end{pmatrix} \cdot \begin{pmatrix} 3 & 4 \\ 2 & 5 \end{pmatrix}$$

A couple of things:

- Unlike numbers, multiplication is not *commutative*, meaning that the order in which you multiply matrices matters a great deal.
- In order to multiply matrices, their dimensions must match up. This means that the first matrix must have the same number of columns as the second matrix has rows.

To remember this, write out the dimensions of the matrices. In this case, we have a 3 x 2 times a 2 x 2 matrix. You can multiple matrices together if the second number in the first dimension pair is the same as the first number in the second dimension pair.

$$3\times\boxed{2\cdot 2}\times 2$$

The resulting matrix will always have dimensions equal to the outer numbers in the dimension pairs (the ones you did not circle in the second point). In this case, the resulting matrix will have a dimension of 3 x 2.

How to multiply matrices

To multiply matrices, there is actually a quite simple procedure. Essentially, we are performing a bunch of dot products.

Recall our earlier sample problem, which was as follows:

$$\begin{pmatrix} 1 & 5 \\ 5 & 8 \\ 7 & 8 \end{pmatrix} \cdot \begin{pmatrix} 3 & 4 \\ 2 & 5 \end{pmatrix}$$

We know that our resulting matrix will have a dimension of 3 x 2. So, we know it will look something like the following:

$$\begin{pmatrix} m_{11} & m_{12} \\ m_{21} & m_{22} \\ m_{31} & m_{32} \end{pmatrix}$$

Note that each element of the matrix is indexed using a double index. The first number represents the row, and the second number represents the column. So, the element m_3 is the element in the third row, second column. Each element is the result of a dot product between rows and columns of the original matrices.

The m_{xy} element is the result of the dot product of the xth row of the first matrix and the yth column of the second matrix. Let's solve a few:

$$m_{11} = \begin{pmatrix} 1 \\ 5 \end{pmatrix} \cdot \begin{pmatrix} 3 \\ 2 \end{pmatrix} = 13$$

$$m_{12} = \begin{pmatrix} 1 \\ 5 \end{pmatrix} \cdot \begin{pmatrix} 4 \\ 5 \end{pmatrix} = 29$$

Moving on, we will eventually get a resulting matrix as follows:

$$\begin{pmatrix} 13 & 29 \\ 31 & 60 \\ 37 & 68 \end{pmatrix}$$

Way to go! Let's come back to the movie recommendation example. Recall the user's movie genre preferences of comedy, romance, and action, which are illustrated as follows:

$$U = user\ prefs = \begin{pmatrix} 5 \\ 1 \\ 3 \end{pmatrix}$$

Now suppose we have 10,000 movies, all with a rating for these three categories. To make a recommendation, we need to take the dot product of the preference vector with each of the 10,000 movies. We can use matrix multiplication to represent this.

Instead of writing them all out, let's express it using the matrix notation. We already have *U*, defined here as the user's preference vector (it can also be thought of as a *3 x 1* matrix), and we also need a movie matrix:

$M = movies = 3x10,000$ dimension matrix.

So, now we have two matrices, one is *3 x 1* and the other is *3 x 10,000*. We can't multiply these matrices as they are because the dimensions do not work out. We will have to change *U* a bit. We can take the *transpose* of the matrix (turning all rows into columns and columns into rows). This will switch the dimensions around:

$$U^T = transpose\,of\,U = (5 1 3)$$

So, now we have two matrices that can be multiplied together. To visualize what this looks like:

$$\underset{1\times 3}{(5 1 3 5 1 3)} \cdot \underset{3\times 10000}{\begin{pmatrix} 4 5 2 & & \\ & & \cdots \\ 1 5 1 & & \end{pmatrix}}$$

The resulting matrix will be a *1 x 1,000* matrix (a vector) of 10,000 predictions for each individual movie. Let's try this out in Python!

```
import numpy as np

# create user preferences
user_pref = np.array([5, 1, 3])

# create a random movie matrix of 10,000 movies
movies = np.random.randint(5,size=(3,1000))+1

# Note that the randint will make random integers from 0-4
# so I added a 1 at the end to increase the scale from 1-5
```

We are using the `numpy` array function to create our matrices. We will have both a `user_pref` and a `movies` matrix to represent our data.

To check our dimensions, we can use the `numpy shape` variable, as shown:

```
print user_pref.shape # (1, 3)

print movies.shape    # (3, 1000)
```

This checks out. Last but not least, let's use the matrix multiplication method of numpy (called dot) to perform the operation, as illustrated:

```
# np.dot does both dot products and matrix multiplication
np.dot(user_pref, movies)
```

The result is an array of integers that represents the recommendations of each movie.

For a quick extension of this, let's run some code that predicts across more than 10,000 movies, as shown:

```
import time

for num_movies in (10000, 100000, 1000000, 10000000, 100000000):
    movies = np.random.randint(5,size=(3, movies))+1
    now = time.time()
    np.dot(user_pref, movies)
    print (time.time() - now), "seconds to run", movies, "movies"

0.000160932540894 seconds to run 10000 movies
0.00121188163757 seconds to run 100000 movies
0.0105860233307 seconds to run 1000000 movies
0.096577167511 seconds to run 10000000 movies
4.16197991371 seconds to run 100000000 movies
```

It took only a bit over 4 seconds to run through 100,000,000 movies using matrix multiplication.

Summary

In this chapter, we took a look at some basic mathematical principles that will become very important as we progress through this book. Between logarithms/exponents, matrix algebra, and proportionality, mathematics clearly has a big role not just in the analysis of data but in many aspects of our lives.

The coming chapters will take a much deeper dive into two big areas of mathematics: probability and statistics. It will become our goal to define and interpret the smallest and biggest theorems in these two giant fields of mathematics.

It is in the next few chapters that everything will start to come together. So far in this book, we have looked at math examples, data exploration guidelines, and basic insights into the types of data. It is time to begin to tie all of these concepts together.

5

Impossible or Improbable – A Gentle Introduction to Probability

Over the next few chapters, we will explore both probability and statistics as methods of examining both data-driven situations and real-world scenarios. The rules of probability govern the basics of prediction. We use probability to define the chances of the occurrence of an event.

In this chapter, we will look at the following topics:

- What is probability?
- The differences between the Frequentist approach and the Bayesian approach
- How to visualize probability
- How to utilize the rules of probability
- Using confusion matrices to look at the basic metrics

Probability will help us model real-life events that include a sense of randomness and chance. Over the next two chapters, we will look at the terminology behind probability theorems and how to apply them to model situations that can appear unexpectedly.

Basic definitions

One of the most basic concepts of probability is the concept of a procedure. A **procedure** is an act that leads to a result. For example, throwing a dice or visiting a website.

An **event** is a collection of the outcomes of a procedure, such as getting a heads on a coin flip or leaving a website after only 4 seconds. A simple event is an outcome/ event of a procedure that cannot be broken down further. For example, rolling two dice can be broken down into two simple events: rolling die 1 and rolling die 2.

The **sample space** of a procedure is the set of all possible simple events. For example, an experiment is performed, in which a coin is flipped three times in succession. What is the size of the sample space for this experiment?

The answer is eight, because the results could be any one of the possibilities in the following sample space—{HHH, HHT, HTT, HTH, TTT, TTH, THH, or THT}.

Probability

The **probability** of an event represents the frequency, or chance, that the event will happen.

For **notation**, if *A* is an event, *P(A)* is the probability of the occurrence of the event.

We can define the actual probability of an event, *A*, as follows:

$$P(A) = \frac{\text{number of ways A occur}}{\text{size of sample space}}$$

Here, *A* is the event in question. Think of an entire universe of events where anything is possible, and let's represent it as a circle. We can think of a single event, **A**, as being a smaller circle within that larger universe, as shown in the following diagram:

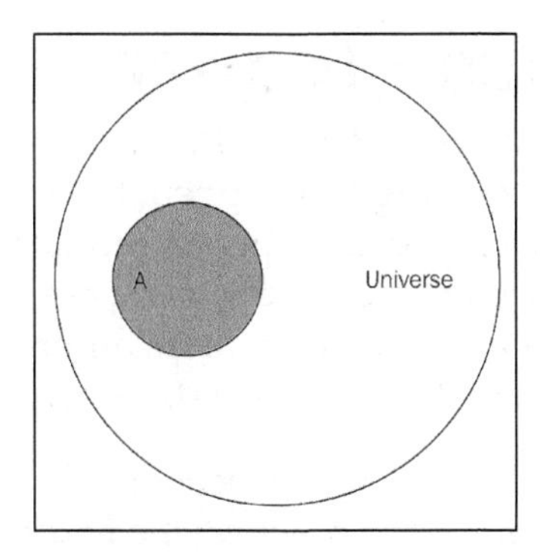

Let's now pretend that our universe involves a research study on humans, and the **A** event is people in that study who have cancer.

If our study has 100 people and *A* has 25 people, the probability of *A* or *P(A)* is 25/100.

The maximum probability of any event is 1. This can be understood as the red circle grows so large that it is the size of the universe (the larger circle).

The most basic examples (I promise they will get more interesting) are coin flips. Let's say we have two coins and we want the probability that we will roll two heads. We can very easily count the number of ways two coins could end up being two heads. There's only one! Both coins have to be heads. But how many options are there? It could either be two heads, two tails, or a heads/tails combination.

First, let's define A. It is the event in which two heads occur. The number of ways that A can occur is 1.

The sample space of the experiment is {HH, HT, TH, TT}, where each two letter *word* indicates the outcome of the first and second coin simultaneously. The size of the sample space is four. So, *P(getting two heads) = 1/4.*

Let's refer to a quick visual table to prove it. The following table denotes the options for coin 1 as the columns and the options for coin 2 as the rows. In each cell, there is either a True or a False. A True value indicates that it satisfies the condition (both heads) and False indicates otherwise.

	Coin 1 is Heads	**Coin 1 is Tails**
Coin 2 is Heads	True	False
Coin 2 is Tails	False	False

So, we have one out of a total of four possible outcomes.

Bayesian versus Frequentist

The preceding example was almost too easy. In practice, we can hardly ever truly count the number of ways something can happen. For example, let's say that we want to know the probability of a random person smoking cigarettes at least once a day. If we wanted to approach this problem using the classical way (the previous formula), we would need to figure out how many different ways a person is a smoker—someone who smokes at least once a day—which is not possible!

When faced with such a problem, two main schools of thought are considered when it comes to calculating probabilities in practice: the **Frequentist approach** and the **Bayesian approach**. This chapter will focus heavily on the Frequentist approach while the subsequent chapter will dive into the Bayesian analysis.

Frequentist approach

In a Frequentist approach, the probability of an event is calculated through experimentation. It uses the past in order to predict the future chance of an event. The basic formula is as follows:

$$P(A) = \frac{\text{number of times A occurred}}{\text{number of times the procedure was repeated}}$$

Basically, we observe several instances of the event and count the number of times *A* was satisfied. The division of these numbers is an approximation of the probability.

The Bayesian approach differs by dictating that probabilities must be discerned using theoretical means. Using the Bayes approach, we would have to think a bit more critically about events and why they occur. Neither methodology is 100% the correct answer all the time. Usually, it comes down to the problem and the difficulty of using either approach.

The crux of the Frequentist approach is the relative frequency.

The **relative frequency** of an event is how often an event occurs divided by the total number of observations.

Example - marketing stats

Let's say that you are interested in ascertaining how often a person who visits your website is likely to return on a later date. This is sometimes called the rate of repeat visitors. In the previous definition, we would define our *A* event as being a visitor coming back to the site. We would then have to calculate the number of ways a person can come back, which doesn't really make sense at all! In this case, many people would turn to a Bayesian approach; however, we can calculate what is known as relative frequency.

So, in this case, we can take the visitor logs and calculate the relative frequency of event *A* (repeat visitors). Let's say, of the *1,458* unique visitors in the past week, *452* were repeat visitors. We can calculate this as follows:

$$P(A)\ RF(A) = \frac{452}{1458} = .31$$

So, about 31% of your visitors are repeat visitors.

The law of large numbers

The reason that even the Frequentist approach can do this is because of the law of large numbers, which states that if we repeat a procedure over and over, the relative frequency probability will approach the actual probability. Let's try to demonstrate this using Python.

If I were to ask you the average of the numbers 1 and 10, you would very quickly answer around 5. This question is identical to asking you to pick the *average* number between 1 and 10. Let's design the experiment to be as follows:

Python will choose *n* random numbers between 1 and 10 and find their average.

We will repeat this experiment several times using a larger *n* each time, and then we will graph the outcome. The steps are as follows:

1. Pick a random number between 1 and 10 and find the average.
2. Pick two random numbers between 1 and 10 and find their average.
3. Pick three random numbers between 1 and 10 and find their average.
4. Pick 10,000 random numbers between 1 and 10 and find their average.
5. Graph the results.

Let's take a look at the code:

```
import numpy as np
import pandas as pd
from matplotlib import pyplot as plt
%matplotlib inline
results = []
for n in range(1,10000):
    nums = np.random.randint(low=1,high=10, size=n) # choose n numbers
between 1 and 10
    mean = nums.mean()                              # find the average
of these numbers
    results.append(mean)                            # add the average
to a running list

# POP QUIZ: How large is the list results?
len(results) # 9999
# This was tricky because I took the range from 1 to 10000 and usually
we do from 0 to 10000
df = pd.DataFrame({ 'means' : results})
print df.head() # the averages in the beginning are all over the
place!
# means
# 9.0
# 5.0
# 6.0
# 4.5
# 4.0
print df.tail() # as n, our size of the sample size, increases, the
averages get closer to 5!
# means
# 4.998799
# 5.060924
# 4.990597
# 5.008802
# 4.979198
df.plot(title='Law of Large Numbers')
plt.xlabel("Number of throws in sample")
plt.ylabel("Average Of Sample")
```

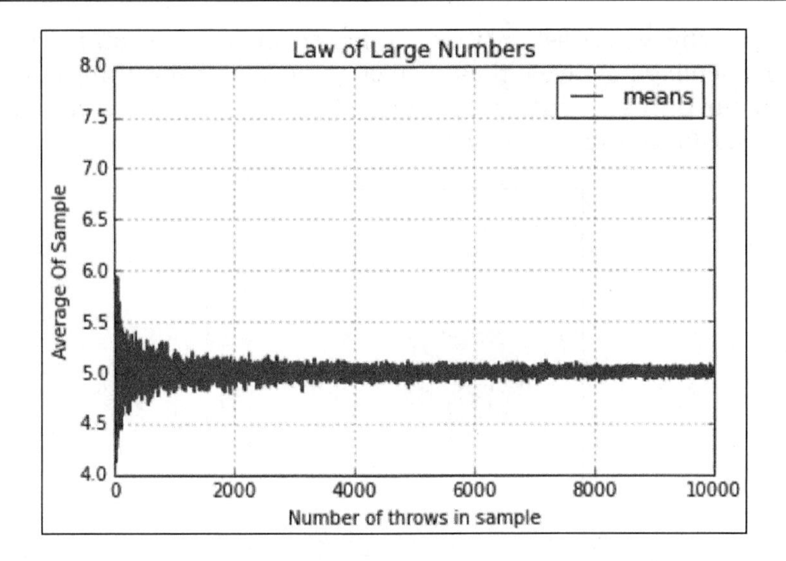

Cool, right? What this is essentially showing us is that as we increase the sample size of our relative frequency, the frequency approaches the actual average (probability) of 5.

In our statistics chapters, we will work to define this law much more rigorously, but for now, just know that it is used to link the relative frequency of an event to its actual probability.

Compound events

Sometimes, we need to deal with two or more events. These are called **compound events**. A compound event is any event that combines two or more simple events. When this happens, we need some special notation.

Given events *A* and *B*:

- The probability that *A* and *B* occur is $P(A \cap B) = P(A \text{ and } B)$
- The probability that either *A* or *B* occurs is $P(A \cup B) = P(A \text{ or } B)$

Understanding why we use set notation for these compound events is very important. Remember how we represented events in a universe using circles earlier? Let's say that our **Universe** is 100 people who showed up for an experiment, in which a new test for cancer is being developed:

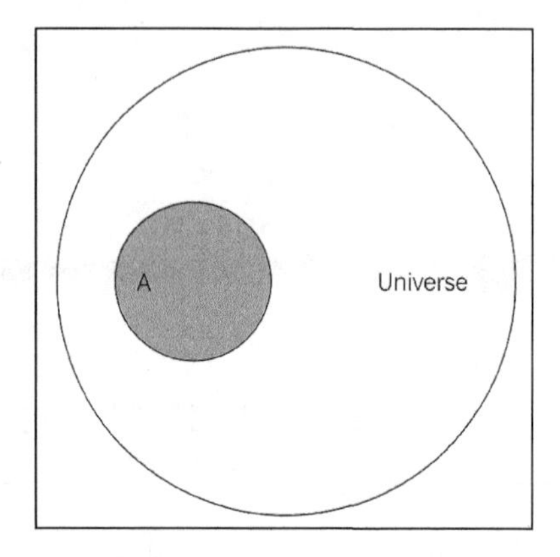

In the preceding diagram, the red circle, **A**, represents 25 people who actually have cancer. Using the relative frequency approach, we can say that *P(A) = number of people with cancer/number of people in study*, that is, *25/100 = ¼ = .25*. This means that there is a 25% chance that someone has cancer.

Let's introduce a second event, called **B**, as shown, which contains people for whom the test was positive (it claimed that they had cancer). Let's say that this is for 30 people. So, *P(B) = 30/100 = 3/10 = .3*. This means that there is a 30% chance that the test said positive for any given person:

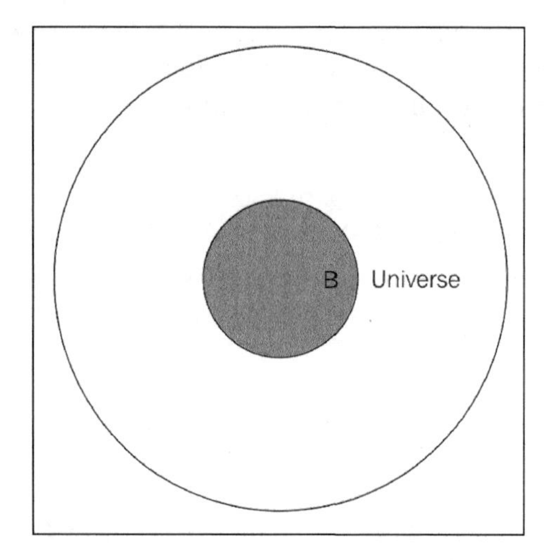

These are two separate events, but they interact with each other. Namely, they might *intersect* or have people in common, as shown here:

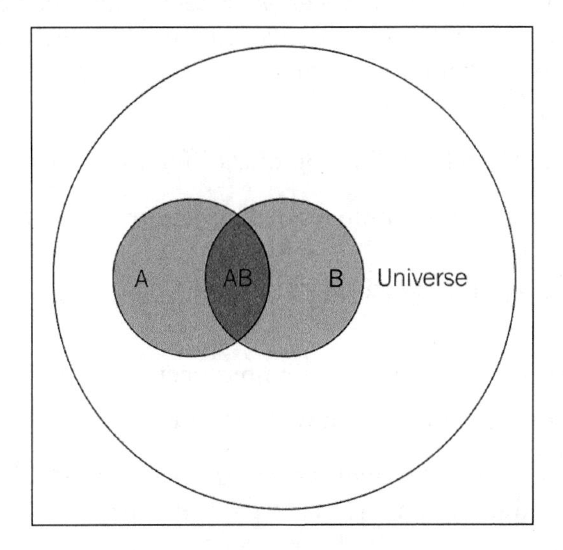

Anyone in the space that both **A** and **B** occupy, otherwise known as *A intersect B* or $A \cap B$, are people for whom the test claimed they were positive for cancer (**A**) and they actually do have cancer. Let's say that's 20 people. The test said positive for 20 people, that is, they have cancer, as shown here:

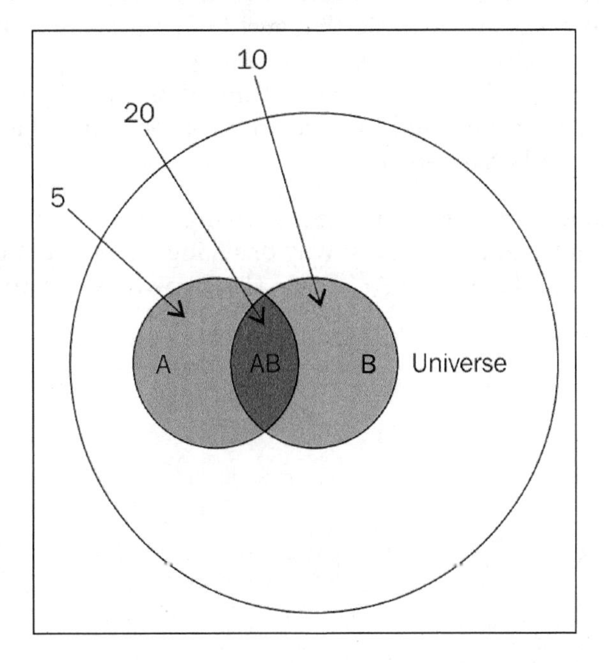

This means that $P(A \text{ and } B) = 20/100 = 1/5 = .2 = 20\%$.

If we want to say that someone has cancer *or* the test came back positive. This would be the total sum (or union) of the two events, namely, the sum of 5, 20, and 10, which is 35. So, 35/100 people either have cancer or had a positive test outcome. That means, $P(A \text{ or } B) = 35/100 = .35 = 35\%$.

All in all, we have people in the following four different classes:

- **Pink**: This refers to the people who have cancer and had a negative test outcome
- **Purple (A intersect B)**: These people have cancer and had a positive test outcome
- **Blue**: This refers to the people with no cancer and a positive test outcome
- **White**: This refers to the people with no cancer and a negative test outcome

So, effectively, the only times the test was *accurate* was in the white and purple regions. In the blue and pink regions, the test was incorrect.

Conditional probability

Let's pick an arbitrary person from this study of 100 people. Let's also assume that you are told that their test result was positive. What is the probability of them actually having cancer? So, we are told that event *B* has already taken place, and that their test came back positive. The question now is: what is the probability that they have cancer, that is *P(A)*? This is called a **conditional probability of A given B** or **P(A | B)**. Effectively, it is asking you to calculate the probability of an event given that another event has already happened.

You can think of conditional probability as changing the relevant universe. *P(A | B)* (called the probability of *A* given *B*) is a way of saying, given that my entire universe is now *B*, what is the probability of *A*? This is also known as transforming the sample space.

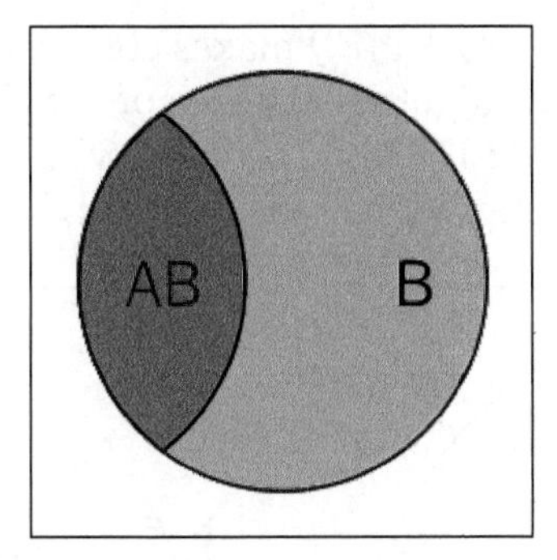

Zooming in on our previous diagram, our universe is now B, and we are concerned with AB (A and B) inside of B

The formula can be given as follows:

$P(A \mid B) = P(A \text{ and } B) / P(B) = (20/100) / (30/100) = 20/30 = .66 = 66\%$

There is a *66%* chance that if a test result came back positive, that person had cancer. In reality, this is the main probability that the experimenters want. They want to know how good the test is at predicting cancer.

The rules of probability

In probability, we have some rules that become very useful when visualization gets too cumbersome. These rules help us calculate compound probabilities with ease.

The addition rule

The addition rule is used to calculate the probability of *either or* events. To calculate $P(A \cup B) = P(A \text{ or } B)$, we use the following formula:

$P(A \cup B) = P(A) + P(B) - P(A \cap B)$

The first part of the formula *(P(A) + P(B))* makes complete sense. To get the union of the two events, we have to add together the area of the circles in the universe. But why the subtraction of *P(A and B)*? This is because when we add the two circles, we are adding the area of intersection twice, as shown in the following diagram:

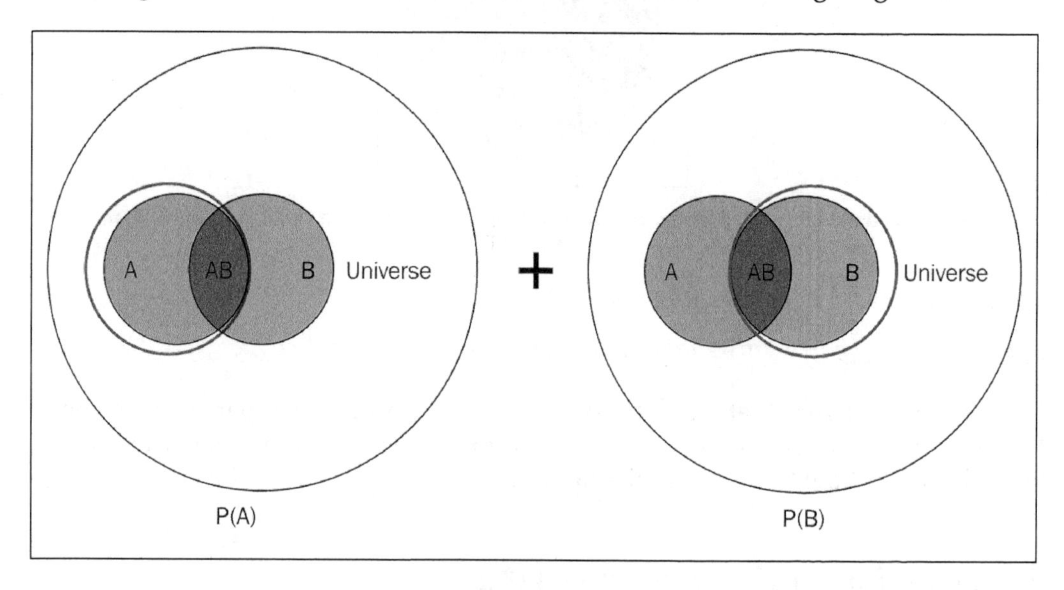

See how both the red circles include the intersection of **A** and **B**? So, when we add them, we need to subtract just one of them to account for this, leaving us with our formula.

Recall that we wanted the number of people who either had cancer or had a positive test result? If *A* is the event that someone has cancer, and *B* is that the test result was positive, we have:

P(A or B) = P(A) + P(B) – P(A and B) = .25 + .30 - .2 = .35

This was calculated before visually in the diagram.

Mutual exclusivity

We say that two events are mutually exclusive if they cannot occur at the same time. This means that $A \cap B = \phi$ or just that the intersection of the events is the empty set. When this happens, *P(A∩B) = P(A and B) = 0.*

If two events are mutually exclusive, then:

$P(A \cup B) = P(A \text{ or } B) = P(A) + P(B) - P(A \cap B) = P(A) + P(B)$

This makes the addition rule much easier. Some examples of mutually exclusive events include the following:

- A customer seeing your site for the first time on both Twitter and Facebook
- Today is Saturday and today is Wednesday
- I failed Econ 101 and I passed Econ 101

None of these events can occur simultaneously.

The multiplication rule

The multiplication rule is used to calculate the probability of *and* events. To calculate $P(A \cap B) = P(A \text{ and } B)$, we use the following formula:

$$P(A \cap B) = P(A \text{ and } B) = P(A) \cdot P(B \mid A)$$

Why do we use $B \mid A$ instead of B? This is because it is possible that B depends on A. If this is the case, then just multiplying $P(A)$ and $P(B)$ does not give us the whole picture.

In our cancer trial example, let's find $P(A \text{ and } B)$. To do this, let's redefine A to be the event that the trial is positive and B to be the person having cancer (because it doesn't matter what we call the events). The equation will be as follows:

$$P(A \cap B) = P(A \text{ and } B) = P(A) \cdot P(B \mid A) = .3 * .6666 = .2 = 20\%$$

This was calculated before visually.

It's difficult to see the true necessity of using the conditional probability, so, let's try another, more difficult problem.

For example, of a randomly selected set of 10 people, 6 have iPhones and 4 have Androids. What is the probability that if I randomly select two people, they both will have iPhones? This example can be retold using event spaces, as follows:

I have the following two events:

- A: This event shows the probability that I choose a person with an iPhone first
- B: This event shows the probability that I choose a person with an iPhone second

So, basically, I want the following:

- *P(A and B)*: *P(I choose a person with an iPhone and a person with an iPhone)*

So, I can use my $P(A \text{ and } B) = P(A) \cdot P(B \mid A)$ formula.

$P(A)$ is simple, right? People with iPhones are 6 out of 10, so, I have a $6/10 = 3/5 = 0.6$ chance of A. This means $P(A) = 0.6$.

So, if I have a 0.6 chance of choosing someone with an iPhone, the probability of choosing two should just be $0.6 * 0.6$, right?

But wait! We only have 9 people left to choose our second person from, because one was taken away. So in our new transformed sample space, we have 9 people in total, 5 with iPhones and 4 with droids, making $P(B) = 5/9 = .555$.

So, the probability of choosing two people with iPhones is $0.6 * 0.555 = 0.333 = 33\%$.

I have a 1/3 chance of choosing two people with iPhones out of 10. The conditional probability is very important in the multiplication rule as it can drastically alter your answer.

Independence

Two events are independent if one event does not affect the outcome of the other, that is $P(B \mid A) = P(B)$ and $P(A \mid B) = P(A)$.

If two events are independent, then:

$$P(A \cap B) = P(A) \cdot P(B \mid A) = P(A) \cdot P(B)$$

Some examples of independent events are as follows:

- It was raining in San Francisco, and a puppy was born in India
- Flip a coin and get heads and flip another coin and get tails

None of these pairs of events affect each other.

Complementary events

The complement of A is the opposite or negation of A. If A is an event, $\overline{A}$ represents the complement of A. For example, if A is the event where someone has cancer, $\overline{A}$ is the event where someone is cancer free.

To calculate the probability of $\overline{A}$, use the following formula:

$P(\overline{A}) = 1 - P(A)$

For example, when you throw two dice, what is the probability that you rolled higher than a 3?

Let A represent rolling higher than a 3.

$\overline{A}$ represents rolling a 3 or less.

$P(A) = 1 - P(\overline{A})$

$P(A) = 1 - (P(2)+P(3))$

$= 1 - (2/36 + 2/36)$

$= 1 - (4/36)$

$= 32/36 = 8/9$

$= .89$

For example, a start-up team has three investor meetings coming up. We will have the following probabilities:

- 60% chance of getting money from the first meeting
- 15% chance of getting money from the second
- 45% chance of getting money from the third

What is the probability of them getting money from at least one meeting?

Let A be the team getting money from at least one investor, and $\overline{A}$ be the team not getting any money. $P(A)$ can be calculated as follows:

$P(A) = 1 - P(\overline{A})$

To calculate $P(\overline{A})$, we need to calculate the following:

$P(\overline{A})$ = *P(no money from investor 1 AND no money from investor 2 AND no money from investor 3)*

If we assume that these events are independent (they don't talk to each other), then:

$P(\bar{A})$ = *P(no money from investor 1) * P(no money from investor 2) * P(no money from investor 3) =*

*0.4 * 0.85 * 0.55 = 0.187*

P(A) = 1 - 0.187 = 0.813 = 81%

So, the startup has an 81% chance of getting money from at least one meeting!

A bit deeper

Without getting too deep into the machine learning terminology, this test is what is known as a **binary classifier**, which means that it is trying to predict from only two options: have cancer or no cancer. When we are dealing with binary classifiers, we can draw what are called confusion matrices, which are 2 x 2 matrices that house all the four possible outcomes of our experiment.

Let's try some different numbers. Let's say 165 people walked in for the study. So, our *n* (sample size) is 165 people. All 165 people are given the test and asked if they have cancer (provided through various other means). The following **confusion matrix** shows us the results of this experiment:

n=165	Predicted: NO	Predicted: YES
Actual: NO	50	10
Actual: YES	5	100

The matrix shows that 50 people were predicted to have no cancer and did not have it, 100 people were predicted to have cancer and actually did have it, and so on. We have the following four classes, again, all with different names:

- The **true positives** are the tests correctly predicting *positive (cancer) == 100*
- The **true negatives** are the tests correctly predicting *negative (no cancer) == 50*
- The **false positives** are the tests incorrectly predicting *positive (cancer) == 10*
- The **false negatives** are the tests incorrectly predicting *negative (no cancer) == 5*

The first two classes indicate where the test was correct or true. The last two classes indicate where the test was incorrect or false.

False positives are sometimes called a **Type I error** whereas false negatives are called a **Type II error**.

Credit: `http://marginalrevolution.com/marginalrevolution/2014/05/type-i-and-type-ii-errors-simplified.html`

We will get into this in the later chapters. For now, we just need to understand why we use the set notation to denote probabilities for compound events. This is because that's what they are. When events A and B exist in the same universe, we can use intersections and unions to represent them happening either at the same time or to represent one happening versus the other.

We will go into this much more in later chapters, but it is good to introduce it now.

Summary

In this chapter, we looked at the basics of probability and will continue to dive deeper into this field in the following chapter. We approached most of our thinking as a Frequentist, and expressed the basics of experimentation and using probability to predict outcome.

The next chapter will look at the Bayesian approach to probability and will also explore the use of probability to solve much more complex problems. We will incorporate these basic probability principles in much more difficult scenarios.

6
Advanced Probability

In the previous chapter, we went over the basics of probability and how we can apply simple theorems to complex tasks. To briefly summarize, probability is the mathematics of modeling events that may or may not occur. We use formulas in order to describe these events and even look at how multiple events can behave together.

In this chapter, we will explore more complicated theorems of probability and how we can use them in a predictive capacity.

Advanced topics, such as **Bayes theorem** and **random variables**, give rise to common machine learning algorithms, such as the **Naïve Bayes algorithm** (also covered in this book). This chapter will focus on some of the more advanced topics in probability theory, including the following topics:

- Exhaustive events
- Bayes theorem
- Basic prediction rules
- Random variables

We have one more definition to look at before we get started (the last one before the fun stuff, I promise). We have to look at *collectively exhaustive events*.

Collectively exhaustive events

When given a set of two or more events, if at least one of the events must occur, then such a set of events is said to be **collectively exhaustive**.

Consider the following examples:

- Given a set of events {`temperature < 60, temperature > 90`}, these events are not collectively exhaustive because there is a third option that is not given in this set of events: The temperature could be between `60` and `90`. However, they are **mutually exhaustive** because both cannot happen at the same time.
- In a dice roll, the set of events of rolling a {`1, 2, 3, 4, 5, or 6`} are **collectively exhaustive** because these are the only possible events, and at least one of them must happen.

Bayesian ideas revisited

In the last chapter, we talked, very briefly, about Bayesian ways of thinking. In short, when speaking about Bayes, you are speaking about the following three things and how they all interact with each other:

- A prior distribution
- A posterior distribution
- A likelihood

Basically, we are concerned with finding the posterior. That's the thing we want to know.

Another way to phrase the Bayesian way of thinking is that data shapes and updates our belief. We have a prior probability, or what we naively think about a hypothesis, and then we have a posterior probability, which is what we think about a hypothesis, given some data.

Bayes theorem

Bayes theorem is the big result of Bayesian inference. Let's see how it even comes about. Recall that we previously defined the following:

- *P(A) = The probability that event A occurs*
- *P(A | B) = The probability that A occurs, given that B occurred*
- *P(A, B) = The probability that A and B occurs*
- *P(A, B) = P(A) * P(B | A)*

That last bullet can be read as *the probability that A and B occur is the probability that A occurs times the probability that B occurred, given that A already occurred.*

It's from that last bullet point that Bayes theorem takes its shape.

We know that:

$P(A, B) = P(A) * P(B \mid A)$

$P(B, A) = P(B) * P(A \mid B)$

$P(A, B) = P(B, A)$

So:

$P(B) * P(A \mid B) = P(A) * P(B \mid A)$

Dividing both sides by *P(B)* gives us Bayes theorem, as shown:

$$P(A \mid B) = \frac{P(A) * P(B \mid A)}{P(B)}$$

You can think of Bayes theorem as follows:

- It is a way to get from *P(A | B)* to *P(B | A)* (if you only have one)
- It is a way to get *P(A | B)* if you already know *P(A)* (without knowing B)

Let's try thinking about Bayes using the terms *hypothesis* and *data*. Suppose H = *your hypothesis about the given data* and D = *the data that you are given*.

Bayes can be interpreted as trying to figure out *P(H | D)* (*the probability that our hypothesis is correct, given the data at hand*).

To use our terminology from before:

$$P(H \mid D) = \frac{P(D \mid H) P(H)}{P(D)}$$

- *P(H)* is the probability of the hypothesis before we observe the data, called the prior probability or just prior
- *P(H | D)* is what we want to compute, the probability of the hypothesis after we observe the data, called the posterior
- *P(D | H)* is the probability of the data under the given hypothesis, called the likelihood
- *P(D)* is the probability of the data under any hypothesis, called the normalizing constant

This concept is not far off from the idea of machine learning and predictive analytics. In many cases, when considering predictive analytics, we use the given data to predict an outcome. Using the current terminology, *H* (*our hypothesis*) can be considered our outcome and *P(H | D)* (the probability that our hypothesis is true, given our data) is another way of saying: *what is the chance that my hypothesis is correct, given the data in front of me?*.

Let's take a look at an example of how we can use Bayes formula at the workplace.

Consider that you have two people in charge of writing blog posts for your company—Lucy and Avinash. From past performances, you have liked 80% of Lucy's work and only 50% of Avinash's work. A new blog post comes to your desk in the morning, but the author isn't mentioned. You love the article. A+. What is the probability that it came from Avinash? Each blogger blogs at a very similar rate.

Before we freak out, let's do what any experienced mathematician (and now you) would do. Let's write out all of our information, as shown:

- *H* = hypothesis = the blog came from Avinash
- *D* = data = you loved the blog post

P(H | D) = the chance that it came from Avinash, given that you loved it

P(D | H) = the chance that you loved it, given that it came from Avinash

P(H) = the chance that an article came from Avinash

P(D) = the chance that you love an article

Note that some of these variables make almost no sense without context. *P(D)*, the probability that you would love any given article put on your desk is a weird concept, but trust me, in the context of Bayes formula, it will be relevant very soon.

Also, note that in the last two items, they assume nothing else. *P(D)* does not assume the origin of the blog post; think of *P(D)* as *if an article was plopped on your desk from some unknown source, what is the chance that you'd like it?* (again, I know it sounds weird out of context).

So, we want to know *P(H | D)*. Let's try to use Bayes theorem, as shown, here:

$$P(H \mid D) = \frac{P(D \mid H) P(H)}{P(D)}$$

But do we know the numbers on the right-hand side of this equation? I claim we do! Let's see here:

- *P(H)* is the probability that any given blog post comes from Avinash. As bloggers write at a very similar rate, we can assume this is .5 because we have a 50/50 chance that it came from either blogger (note how I did not assume D, the data, for this).
- *P(D | H)* is the probability that you love a post from Avinash, which we previously said was 50%, so, .5.
- *P(D)* is interesting. This is the chance that you love an article *in general*. It means that we must take into account the scenario if the post came from Lucy or Avinash. Now, if the hypothesis forms a suite, then we can use our laws of probability, as mentioned in the previous chapter. A suite is formed when a set of hypotheses is both collectively exhaustive and mutually exclusive. In laymen's terms, in a suite of events, exactly one and only one hypothesis can occur. In our case, the two hypotheses are that the article came from Lucy, or that the article came from Avinash. This is definitely a suite because of the following reasons:
    - At least one of them wrote it
    - At most one of them wrote it
    - Therefore, *exactly one* of them wrote it

When we have a suite, we can use our multiplication and addition rules, as follows:

$$D = (From\ Avinash\ AND\ loved\ it) \quad OR \quad (From\ Lucy\ AND\ loved\ it)$$

$$P(D) = P(Loved\ AND\ from\ Avinash) \quad OR \quad P(Loved\ AND\ from\ Lucy)$$

$$P(D) = P(From\ Avinash)P(Loved \mid form\ Avinash) + P(from\ Lucy)P(Loved \mid from\ Lucy)$$

$$P(D) = .5(.5) + .5(.8) = \mathbf{.65}$$

Whew! Way to go. Now we can finish our equation, as shown:

$$P(H \mid D) = \frac{P(D \mid H)P(H)}{P(D)}$$

$$P(H \mid D) = \frac{.5 * .5}{.65} = .38$$

This means that there is a 38% chance that this article comes from Avinash. What is interesting is that *P(H) = .5* and *P(H | D) = .38*. It means that without any data, the chance that a blog post came from Avinash was a coin flip, or 50/50. Given some data (your thoughts on the article), we updated our beliefs about the hypothesis and it actually lowered the chance. This is what Bayesian thinking is all about—updating our posterior beliefs about something from a prior assumption, given some new data about the subject.

More applications of Bayes theorem

Bayes theorem shows up in a lot of applications, usually when we need to make fast decisions based on data and probability. Most recommendation engines, such as Netflix's, use some elements of Bayesian updating. And if you think through why that might be, it makes sense.

Let's suppose that in our simplistic world, Netflix only has 10 categories to choose from. Now suppose that given no data, a user's chance of liking a comedy movie out of 10 categories is 10% (just 1/10).

Okay, now suppose that the user has given a few comedy movies 5/5 stars. Now when Netflix is wondering what the chance is that the user would like another comedy, the probability that they might like a comedy, *P(H | D)*, is going to be larger than a random guess of 10%!

Let's try some more examples of applying Bayes theorem using more data. This time, let's get a bit grittier.

Example – Titanic

A very famous dataset involves looking at the survivors of the sinking of the Titanic in 1912. We will use an application of probability in order to figure out if there were any demographic features that showed a relationship to passenger survival. Mainly, we are curious to see if we can isolate any features of our dataset that can tell us more about the types of people who were likely to survive this disaster.

First, let's read in the data, as shown here:

```
titanic = pd.read_csv(data/titanic.csv')#read in a csv
titanic = titanic[['Sex', 'Survived']]  #the Sex and Survived column
titanic.head()
```

	Sex	Survived
0	male	no
1	female	yes
2	female	yes
3	female	yes
4	male	no

In the preceding table, each row represents a single passenger on the ship, and, for now, we are looking at two specific features: the sex of the individual and whether or not they survived the sinking. For example, the first row represents a man who did not survive while the fourth row (with index 3, remember how python indexes lists) represents a female who did survive.

Let's start with some basics. Let's start by calculating the probability that any given person on the ship survived, regardless of their gender. To do this, let's count the number of yeses in the Survived column and divide this figure by the total number of rows, as shown here:

```
num_rows = float(titanic.shape[0]) # == 891 rows
p_survived = (titanic.Survived=="yes").sum() / num_rows # == .38
p_notsurvived = 1 - p_survived                          # == .61
```

Note that I only had to calculate *P(Survived)*, and I used the law of conjugate probabilities to calculate *P(Died)* because those two events are complementary. Now, let's calculate the probability that any single passenger is male or female:

```
p_male = (titanic.Sex=="male").sum() / num_rows     # == .65
p_female = 1 - p_male # == .35
```

Now let's ask ourselves a question, did having a certain gender affect the survival rate? For this, we can estimate *P(Survived | Female)* or the chance that someone survived given that they were a female. For this, we need to divide the number of women who survived by the total number of women, as shown here:

$$P(Survived \mid Female) - \frac{P(Female\ AND\ Survived)}{P(Female)}$$

```
number_of_women = titanic[titanic.Sex=='female'].shape[0] # == 314
women_who_lived = titanic[(titanic.Sex=='female') & (titanic.
Survived=='yes')].shape[0]                        # == 233
p_survived_given_woman = women_who_lived / float(number_of_women)
p_survived_given_woman              # == .74
```

That's a pretty big difference. It seems that gender plays a big part in this dataset.

Example – medical studies

A classic use of Bayes theorem is the interpretation of medical trials. Routine testing for illegal drug use is increasingly common in workplaces and schools. The companies that perform these tests maintain that the tests have a high sensitivity, which means that they are likely to produce a positive result *if there are drugs in their system*. They claim that these tests are also highly specific, which means that they are likely to yield a negative result *if there are no drugs*.

On average, let's assume that the sensitivity of common drug tests is about 60% and the specificity is about 99%. It means that if an employee is using drugs, the test has a 60% chance of being positive, while if an employee is not on drugs, the test has a 99% chance of being negative. Now, suppose these tests are applied to a workforce where the actual rate of drug use is 5%.

The real question here is *of the people who test positive, how many actually use drugs?*

In Bayesian terms, we want to compute the probability of drug use, given a positive test.

Let D = the event that drugs are in use

Let E = the event that the test is positive

Let N = the event that drugs are *NOT* in use

We are looking for $P(D \mid E)$.

By using Bayes theorem , we can extrapolate it as follows:

$$P(D \mid E) = \frac{P(E \mid D)P(D)}{P(E)}$$

The prior, $P(D)$ is the probability of drug use before we see the outcome of the test, which is 5%. The likelihood, $P(E \mid D)$, is the probability of a positive test assuming drug use, which is the same thing as the sensitivity of the test. The normalizing constant, $P(E)$, is a little bit trickier.

We have to consider two things: *P(E and D)* as well as *P(E and N)*. Basically, we must assume that the test is capable of being incorrect when the user is not using drugs. Check out the following equations:

$$P(E) = P(E \, and \, D) \, or \, P(E \, and \, N)$$

$$P(E) = P(D)P(E \mid D) + P(N)P(E \mid N)$$

$$P(E) = .05 * .6 + .95 * .01$$

$$P(E) = 0.0395$$

So, our original equation becomes as follows:

$$P(D \mid E) = \frac{.6 * .05}{0.0395}$$

$$P(D \mid E) = .76$$

This means that of the people who test positive for drug use, about a quarter are innocent!

Random variables

A **random variable** uses real numerical values to describe a probabilistic event. In our previous work with variables (both in math and programming), we were used to the fact that a variable takes on a certain value. For example, we might have a triangle in which we are given a variable *h* for the hypotenuse, and we must figure out the length of the hypotenuse. We also might have, in Python:

```
x = 5
```

Both of these variables are equal to one value at a time. In a random variable, we are subject to randomness, which means that our variables' values are, well just that, variable! They might take on multiple values depending on the environment.

A random variable still, as shown previously, holds a value. The main distinction between variables as we have seen them and a random variable is the fact that a random variable's value may change depending on the situation.

However, if a random variable can have many values, how do we keep track of them all? Each value that a random variable might take on is associated with a percentage. For every value that a random variable might take on, there is a single probability that the variable will be this value.

With a random variable, we can also obtain our probability distribution of a random variable, which gives the variable's possible values and their probabilities.

Written out, we generally use single capital letters (mostly the specific letter *X*) to denote random variables. For example, we might have:

- *X* = the outcome of a dice roll
- *Y* = the revenue earned by a company this year
- *Z* = the score of an applicant on an interview coding quiz (0-100%)

Effectively, a random variable is a function that maps values from the sample space of an event (the set of all possible outcomes) to a probability value (between 0 and 1). Think about the event as being expressed as the following:

$$f(event) = probability$$

It will assign a probability to each individual option. There are two main types of random variables: discrete and continuous.

Discrete random variables

A discrete random variable only takes on a countable number of possible values. For example, the outcome of a dice roll, as shown here:

$$\text{X = the outcome of a single dice roll}$$

Value	$X = 1$	$X = 2$	$X = 3$	$X = 4$	$X = 5$	$X = 6$
Probability	$\frac{1}{6}$	$\frac{1}{6}$	$\frac{1}{6}$	$\frac{1}{6}$	$\frac{1}{6}$	$\frac{1}{6}$

Note how I use a capital X to define the random variable. This is a common practice.

Also, note how the random variable maps a probability to each individual outcome.

Random variables have many properties, two of which are their *expected value* and the *variance*.

We will use a **probability mass function** (**PMF**) to describe a discrete random variable.

They take on the appearance of the following:

P(X = x) = PMF

So, for a dice roll, *P(X = 1) = 1/6* and *P(X = 5) = 1/6*.

Consider the following examples of discrete variables:

- The likely result of a survey question (for example, on a scale of 1-10)
- Whether the CEO will resign within the year (either true or false)

The expected value of a random variable defines the mean value of a long run of repeated samples of the random variable. This is sometimes called the mean of the variable.

For example, refer to the following Python code that defines the random variable of a dice roll:

```
import random
def random_variable_of_dice_roll():
    return random.randint(1, 7) # a range of (1,7) # includes 1, 2, 3,
4, 5, 6, but NOT 7
```

This function will invoke a random variable and come out with a response. Let's roll `100` dice and average the result, as follows:

```
trials = []
num_trials = 100
for trial in range(num_trials):
trials.append( random_variable_of_dice_roll() )
print sum(trials)/float(num_trials)   # == 3.77
```

So, taking `100` dice rolls and averaging them gives us a value of `3.77`! Let's try this with a wide variety of trial numbers, as illustrated here:

```
num_trials = range(100,10000, 10)
avgs = []
for num_trial in num_trials:
    trials = []
    for trial in range(1,num_trial):
```

```
        trials.append( random_variable_of_dice_roll() )
    avgs.append(sum(trials)/float(num_trial))

plt.plot(num_trials, avgs)
plt.xlabel('Number of Trials')
plt.ylabel("Average")
```

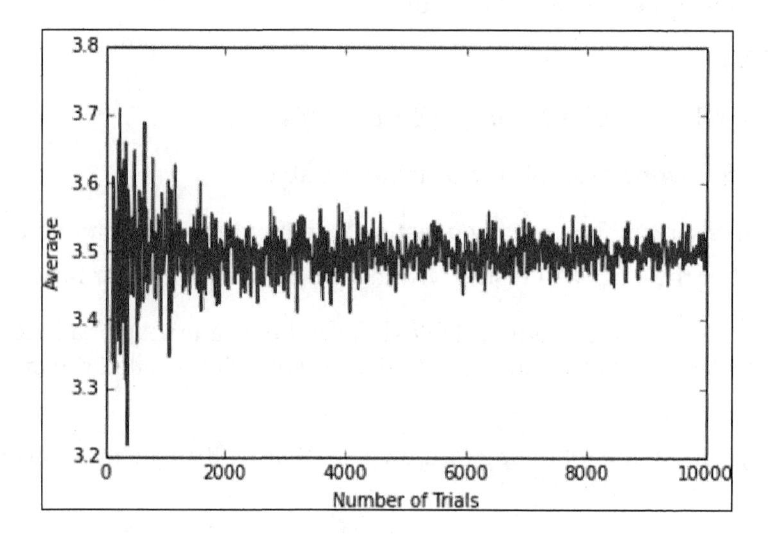

The preceding graph represents the average dice roll as we look at more and more dice rolls. We can see that the average dice roll is rapidly approaching 3.5. If we look towards the left of the graph, we see that if we only roll a die about 100 times, then we are not guaranteed to get an average dice roll of 3.5. However, if we roll 10,000 dice one after another, we see that we would very likely expect the average dice roll to be about 3.5.

For a discrete random variable, we can also use a simple formula, shown as follows, to calculate the expected value:

$$\text{Expected value} = E[X] = \mu_X = \sum x_i p_i$$

Where x_i is the ith outcome and p_i is the ith probability.

So, for our dice roll, we can find the exact expected value as being as follows:

$$\frac{1}{6}(1)+\frac{1}{6}(2)+\frac{1}{6}(3)+\frac{1}{6}(4)+\frac{1}{6}(5)+\frac{1}{6}(6)=3.5$$

The preceding result shows us that for any given dice roll, we can "expect" a dice roll of 3.5. Now, obviously, that doesn't make sense because we can't get a 3.5 on a dice roll, but it does make sense when put in the context of many dice rolls. If you roll 10,000 dice, your average dice roll should approach 3.5, as shown in the graph and code previously.

The average of the expected value of a random variable is generally not enough to grasp the full idea behind the variable. For this reason, we introduce a new concept, called variance.

The variance of a random variable represents the spread of the variable. It quantifies the variability of the expected value.

The formula for the variance of a discrete random variable is expressed as follows:

$$\text{Variance} = V[X] = \sigma_X^2 = \sum (x_i - \mu_X)^2 p_i$$

Where x_i and p_i represent the same values as before and μ_x represents the expected value of the variable. In this formula, I also mentioned sigma of X. Sigma, in this case, is the standard deviation, which is defined simply as the square root of the variance. Let's look at a more complicated example of a discrete random variable.

Variance can be thought of as a *give or take* metric. If I say you can *expect* to win $100 off of a poker hand, you might be very happy. If I append that statement with the additional detail that you might win $100, give or take $80, you now have a wide range of expectations to deal with, which can be frustrating and might make a risk-averse player more wary of joining the game. We can usually say that we have an expected value, give or take the standard deviation.

Consider that your team measures the success of a new product on a **likert scale**, that is, as being in one of five categories, where a value of 0 represents a complete failure and 4 represents a great success. They estimate that a new project has the following chances of success based on user testing and the preliminary results of the performance of the product.

We first have to define our random variable.

Let the *X* random variable represent the success of our product. *X* is indeed a discrete random variable because the *X* variable can only take on one of five options: *0, 1, 2, 3,* or *4*.

The following is the probability distribution of our random variable, X. Note how we have a column for each potential outcome of X and below each outcome we have the probability that that particular outcome will be achieved:

Value	$X = 0$	$X = 1$	$X = 2$	$X = 3$	$X = 4$
Probability	0.02	0.07	0.25	0.4	0.26

For example, the project has a 2% chance of failing completely and a 26% chance of being a great success! We can calculate our expected value as follows:

$E[X] = 0(0.02) + 1(0.07) + 2(0.25) + 3(0.4) + 4(0.26) = 2.81$

This number means that the manager can *expect* a success of about 2.81 out of this project. Now, by itself, that number is not very useful. Perhaps, if given several products to choose from, an expected value might be a way to compare the potential successes of several products. However, in this case, when we have but the one product to evaluate, we will need more.

Now, let's check the variance, as shown here:

$Variance=V[X]=\sigma X2 = (xi - \mu X)2pi$

$= (0 - 2.81)2(0.02) + (1 - 2.81)2(0.07)+ (2 - 2.81)2(0.25) + (3 - 2.81)2(0.4) + (4 - 2.81)2(0.26) = .93$

Now that we have both the standard deviation and the expected value of the score of the project, let's try to summarize our results. We could say that our project will have an expected score of 2.81 plus or minus .96 meaning that can expect something between 1.85 and 3.77.

So, one way we can address this project is that it is probably going to have a success rating of *2.81*, give or take about a point.

You might be thinking, *wow, Sinan, so, at best the project will be a 3.8 and at worst it will be a 1.8?*. Not quite.

It might be better than a 4 and it might also be worse than a 1.8. To take this one step further, let's calculate the following:

$P(X >= 3)$

First, take a minute and convince yourself that you can read that formula to yourself. What am I asking when I am asking for *P(X >= 3)*? Honestly, take a minute and figure it out.

P(X >= 3) is the probability that our random variable will take on a value at least as big as 3. In other words, what is the chance that our product will have a success rating of 3 or higher? To calculate this, we can calculate the following:

$P(X >= 3) = P(X = 3) + P(X = 4) = .66 = 66\%$

This means that we have a 66% chance that our product will rate as either a 3 or a 4.

Another way to calculate this would be the conjugate way, as shown here:

$P(X >= 3) = 1 - P(X < 3)$

Again, take a moment to convince yourself that this formula holds up. I am claiming that to find the probability that the product will be rated at least a 3 is the same as 1 minus the probability that the product will receive a rating below 3. If this is true, then the two events (*X >=3* and *X < 3*) must complement one another.

This is obviously true! The product can be either of the following two options:

- Be rated 3 or above
- Be rated below a 3

Let's check our math:

$P(X < 3) = P(X = 0) + P(X = 1) + P(X = 2) = 0.02 + 0.07 + 0.25 = .034$

$1 - P(X < 3) = 1 - .34 = .66 = P(x >= 3)$

It checks out!

Types of discrete random variables

We can get a better idea of how random variables work in practice by looking at specific types of random variables. These specific types of random variables model different types of situations and end up revealing much simpler calculations for very complex event modeling.

Binomial random variables

The first type of discrete random variable we will look at is called a **binomial random variable**. With a binomial random variable, we look at a setting in which a single event happens over and over and we try to count the number of times the result is positive.

Before we can understand the random variable itself, we must look at the conditions in which it is even appropriate.

A binomial setting has the following four conditions:

- The possible outcomes are either success or failure
- The outcomes of trials cannot affect the outcome of another trial
- The number of trials was set (a fixed sample size)
- The chance of success of each trial must always be p

A binomial random variable is a discrete random variable, X, that counts the number of successes in a binomial setting. The parameters are n = *the number of trials* and p = *the chance of success of each trial.*

Example - fundraising meetings:

A start-up is taking 20 VC meetings to fund and count the number of offers they receive.

The **probability mass function (PMF)** for a binomial random variable is as follows:

$$P(X = k) = \binom{n}{k} p^k (1-p)^{n-k}$$

Here, $\binom{n}{k} = the\, binomial\, coefficient = \frac{n!}{(n-k)!k!}$.

Example – restaurant openings

A new restaurant in a town has a 20% chance of surviving its first year. If 14 restaurants open this year, find the probability that exactly four restaurants survive their first year of being open to the public.

First, we should prove that this is a binomial setting:

- The possible outcomes are either success or failure (the restaurants either survive or not)
- The outcomes of trials cannot affect the outcome of another trial (assume that the opening of one restaurant doesn't affect another restaurant's opening and survival)
- The number of trials was set (14 restaurants opened)
- The chance of success of each trial must always be p (we assume that it is always 20%)

Here, we have our two parameters of $n = 14$ and $p = .2$. So, we can now plug these numbers into our binomial formula, as shown here:

$$P(X=4)=\binom{14}{4}.2^4.8^{10}=.17$$

So, we have a 17% chance that exactly 4 of these restaurants will be open after a year.

Example - blood types

A couple has a 25% chance of a having a child with type O blood. What is the chance that 3 of their 5 kids have type O blood?

Let X = *the number of children with type O blood* with $n = 5$ and $p = 0.25$, as shown here:

P(X = 3) = 5 0.253(0.75)5−3 = 10(0.25)3(0.75)2 = 0.087

We can calculate this probability for the values of 0, 1, 2, 3, 4, and 5 to get a sense of the probability distribution:

value x_i	0	1	2	3	4	5
Probability	0.23730	0.39551	0.26367	0.08789	0.01465	0.00098

From here, we can calculate an expected value and the variance of this variable:

$$\text{Expected value} = E[X] = \mu_X = \sum x_i p_i = 1.25$$

$$\text{Variance} = V[X] = \sigma_X^2 = \sum (x_i - \mu_X)^2 p_i = 0.9375$$

So, this family can expect to have probably 1 or 2 kids with type O blood!

What if we want to know the probability that at least 3 of their kids have type O blood? To know the probability that at least three of their kids have type O blood, we can use the following formula for discrete random variables:

$$P(X \geq 3) = P(X=5) + P(X=4) + P(X=3)$$

$$= .00098 + .01465 + .08789 = 0.103$$

So, there is about a 10% chance that three of their kids have type O blood.

Shortcuts to binomial expected value and variance

Binomial random variables have special calculations for the exact values of the expected values and variance. If X is a binomial random variable, then:

$E(X) = np$

$V(X) = np(1 - p)$

For our preceding example, we can use the following formulas to calculate an exact expected value and variance:

- $E(X) = .25(5) = 1.25$
- $V(X) = 1.25(.75) = .9375$

A binomial random variable is a discrete random variable that counts the number of successes in a binomial setting. It is used in a wide variety of data-driven experiments, such as counting the number of people who will sign up for a website given a chance of conversion, or even, at a simple level, predicting stock price movements given a chance of decline (don't worry; we will be applying much more sophisticated models to predict the stock market later).

Geometric random variables;

The second discrete random variable we will take a look at is called a geometric random variable. It is actually quite similar to the binomial random variable in that we are concerned with a setting in which a single event is occurring over and over. However, in the case of a geometric setting, the major difference is that we are not fixing the sample size.

We are not going into exactly 20 VC meetings as a start-up, nor are we having exactly 5 kids. Instead, in a geometric setting, we are modeling the number of trials we will need to see before we obtain even a single success. Specifically, a geometric setting has the following four conditions:

- The possible outcomes are either success or failure
- The outcomes of trials cannot affect the outcome of another trial
- The number of trials was not set
- The chance of success of each trial must always be p

Note that these are the exact same conditions as a binomial variable, except the third condition.

A **geometric random variable** is a discrete random variable, X, that counts the number of trials needed to obtain one success. The parameters are *p = the chance of success of each trial* and *(1 − p) = the chance of failure of each trial.*

To transform the previous binomial examples into geometric examples, we might do the following:

- Count the number of VC meetings that a start-up must take in order to get their first *yes*
- Count the number of coin flips needed in order to get a heads (yes, I know it's boring, but it's a solid example!)

The formula for the PMF is as follows:

P(X = x) = (1−p)[x−1]p

Both the binomial and geometric settings involve outcomes that are either successes or failures. The big difference is that binomial random variables have a fixed number of trials, denoted as *n*. Geometric random variables do not have a fixed number of trials. Instead, geometric random variables model the number of samples needed in order to obtain the first successful trial, whatever success might mean in those experimental conditions.

Example – weather

There is a 34% chance that it will rain on any day in April. Find the probability that the first day of rain in April will occur on April fourth.

Let X = *the number of days until it rains* (success) with *p = 0.34* and *(1 − p) = 0.66*

So, *P(X = 8) = (0.66)8−1(0.34)*

= (0.66)7(0.34)

= 0.01855

The probability that it will rain by the fourth of April is as follows:

$$P(X \leq 4) = P(1) + P(2) + P(3) + P(4) =$$

$$= .34 + .22 + .14 + .1 = .8$$

So, there is an 80% chance that the first rain of the month will happen within the first four days.

Shortcuts to geometric expected value and variance

Geometric random variables also have special calculations for the exact values of the expected values and variance. If *X* is a geometric random variable, then,

E(X) = 1/p

V(X) = (1-p)/p2

Poisson random variable,

The third and last specific example of a discrete random variable is a Poisson random variable.

To understand why we would need this random variable, imagine that an event that we wish to model has a small probability of happening and that we wish to count the number of times that the event occurs in a certain time frame. If we have an idea of the average number of occurrences, μ, over a specific period of time, given from past instances, then the Poisson random variable, denoted by $X = Poi(\mu)$, counts the total number of occurrences of the event during that given time period.

In other words, the Poisson distribution is a discrete probability distribution that counts the number of events that occur in a given interval of time.

Consider the following examples of Poisson random variables:

- Finding the probability of having a certain number of visitors on your site within an hour, knowing the past performance of the site
- Estimating the number of car crashes at an intersection based on past police reports

If we let X = *the number of events in a given interval*, and the average number of events per interval is the λ number, then the probability of observing *x* events in a given interval is given by the following formula:

$$P(X = x) = \frac{e^{-\lambda}\lambda^{x}}{x!}$$

Here, *e* = *Euler's constant (2.718....)*.

Example - call center:

The number of calls arriving at your call center follows a Poisson distribution at the rate of 5 calls/hour. What is the probability that exactly six calls will come in between 10 and 11 p.m.?

To set up this example, let's write out our given information. Let *X* be the number of calls that arrive between 10 and 11 p.m. This is our Poisson random variable with mean $\lambda = 5$.

The mean is 5 because we are using 5 as our previous expected value of the number of calls to come in at this time. This number could have come from precious work on estimating the number of calls that come in every hour or specifically that come in after 10 p.m. The main idea is that we do have some idea of how many calls should be coming in, and then we use that information to create our Poisson random variable and use it to make predictions.

Continuing with our example, we have the following:

P(X = 6) = = 0.146

This means that there is about a 14.6% chance that exactly six calls will come between 10 and 11 p.m.

Shortcuts to Poisson expected value and variance

Poisson random variables also have special calculations for the exact values of the expected values and variance. If *X* is a Poisson random variable with mean, then:

$$E(X) = \lambda$$

$$V(X) = \lambda$$

This is actually interesting because both the expected value and the variance are the same number and that number is simply the given parameter! Now that we've seen three examples of discrete random variables, we must take a look at the other type of random variable, called the continuous random variable.

Continuous random variables

Switching gears entirely, unlike a discrete random variable, a continuous random variable can take on an *infinite* number of possible values, not just a few countable ones. We call the functions that describe the distribution density curves instead of probability mass functions.

Consider the following examples of continuous variables:

- The length of a sales representative's phone call (not the number of calls)
- The actual amount of oil in a drum marked 20 gallons (not the number of oil drums)

If X is a continuous random variable, then there is a function, $f(x)$, such that for any constants a and b:

$$P(a \le X \le b) = \int_a^b f(x)\,dx$$

The preceding $f(x)$ function is known as the **probability density function** (**PDF**). The PDF is the continuous random variable version of the PMF for discrete random variables.

The most important continuous distribution is the **standard normal distribution**. You have, no doubt, either heard of the normal distribution or dealt with it. The idea behind it is quite simple. The PDF of this distribution is as follows:

$$f(x) = \frac{1}{\sqrt{2\pi\sigma^2}} e^{-\frac{(x-\mu)^2}{2\sigma^2}}$$

Here, μ is the mean of the variable and σ is the standard deviation. This might look confusing, but let's graph it in Python with a mean of 0 and a standard deviation of 1, as shown here:

```
def normal_pdf(x, mu = 0, sigma = 1):
    return (1./np.sqrt(2*3.14 * sigma**2)) * 2.718**(-(x-mu)**2 / (2.
* sigma**2))

x_values = np.linspace(-5,5,100)
y_values = [normal_pdf(x) for x in x_values]
plt.plot(x_values, y_values)
```

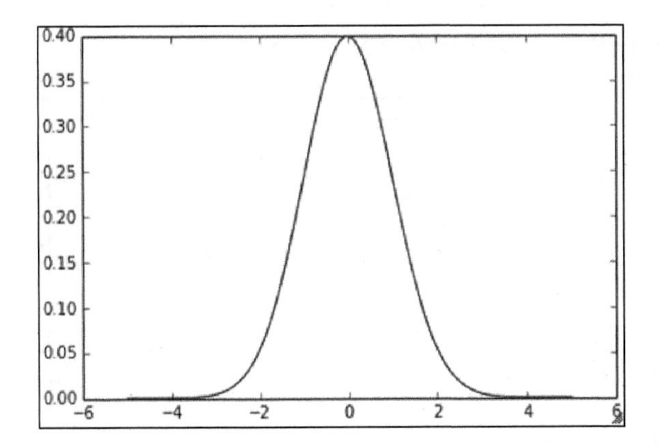

Which gives rise to the all-too-familiar bell curve. Note that the graph is symmetrical around the x = 0 line. Let's try changing some of the parameters. First, let's try with $\mu = 5$:

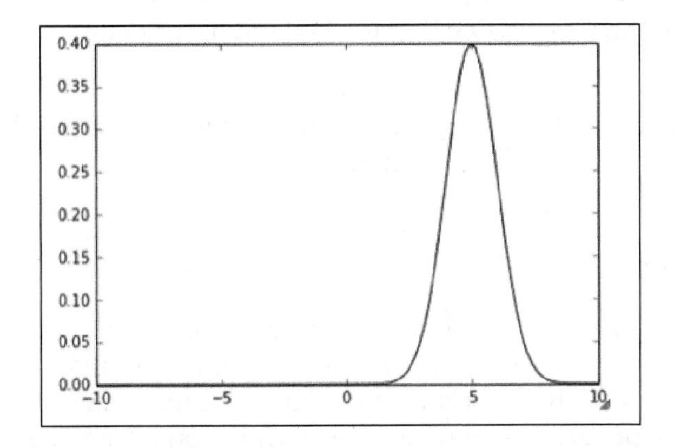

Next, let's try with the value $\sigma = 5$:

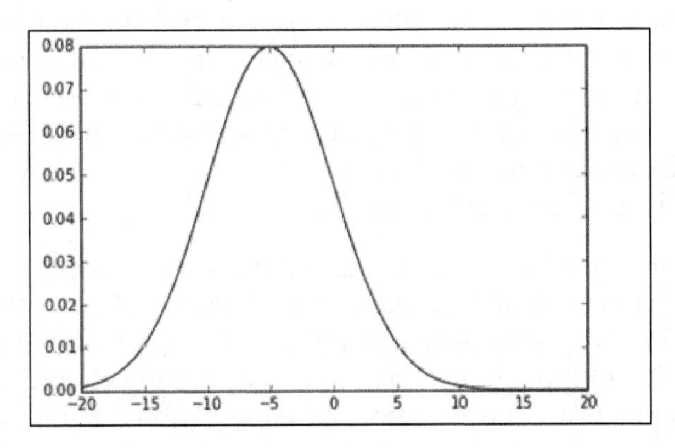

Lastly, we will try with the values $\mu = 5 \ \sigma = 5$:

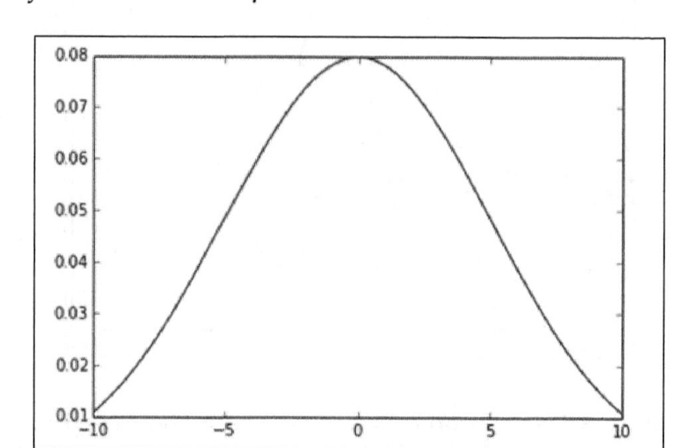

In all the graphs, we have the standard bell shape that we are all familiar with, but as we change our parameters, we see that the bell might get skinnier, thicker, or move from left to right.

In the following chapters that focus on statistics, we will make much more use of the normal distribution as it applies to statistical thinking.

Summary

Probability as a field works to explain our random and chaotic world. Using the basic laws of probability, we can model real-life events that involve randomness. We can use random variables to represent values that may take on several values, and we can use the probability mass or density functions to compare product lines or look at the test results.

We have seen some of the more complicated uses of probability in prediction. Using random variables and Bayes theorem are excellent ways to assign probabilities to real-life situations. In the later chapters, we will revisit Bayes theorem and use it to create a very powerful and fast machine learning algorithm, called Naïve Bayes algorithm. This algorithm captures the power of Bayesian thinking and applies it directly to the problem of predictive learning.

The next two chapters are focused on statistical thinking. Like probability, these chapters will use mathematical formulas to model real-world events. The main difference, however, will be the terminology we use to describe the world and the way we model different types of events. In these upcoming chapters, we will attempt to model entire populations of data points based solely on a sample.

We will revisit many concepts in probability to make sense of statistical theorems as they are closely linked and both are important mathematical concepts in the realm of data science.

7
Basic Statistics

This chapter will focus on the statistics required by any aspiring data scientist.

We will explore ways of sampling and obtaining data without being affected by bias and then use measures of statistics to quantify and visualize our data. Using the z-score and the Empirical rule, we will see how we can standardize data for the purpose of both graphing and interpretability.

In this chapter, we will look at the following topics:

- How to obtain and sample data
- The measures of center, variance, and relative standing
- Normalization of data using the z-score
- The Empirical rule

What are statistics?

This might seem like an odd question to ask, but I am frequently surprised by the number of people who cannot answer this simple and yet powerful question: what are statistics? Statistics are the numbers you always see on the news and in the paper. Statistics are useful when trying to prove a point or trying to scare you, but what are they?

To answer this question, we need to back up for a minute and talk about why we even measure them in the first place. The goal of this field is to try to explain and model the world around us. To do that, we have to take a look at the population.

We can define a **population** as the entire pool of subjects of an experiment or a model.

Essentially, your population is who you care about. Who are you trying to talk about? If you are trying to test if smoking leads to heart disease, your population would be the smokers of the world. If you are trying to study teenage drinking problems, your population would be all teenagers.

Now, consider that you want to ask a question about your population, for example, if your population is all of your employees (assume that you have over 1,000 employees), perhaps you want to know what percentage of them use illicit drugs. The question is called a **parameter**.

We can define a parameter as a numerical measurement describing a characteristic of a population.

For example, if you ask all 1,000 employees and 100 of them are using drugs, the rate of drug use is 10%. The parameter here is 10%.

However, let's get real, you probably can't ask every single employee whether they are using drugs. What if you have over 10,000 employees? It would be very difficult to track everyone down in order to get your answer. When this happens, it's impossible to figure out this parameter. In this case, we can *estimate* the parameter.

First, we will take a **sample** of the population.

We can define a sample of a population as a subset (random not required) of the population.

So, we perhaps ask 200 of the 1,000 employees you have. Of these 200, suppose 26 use drugs, making the drug use rate 13%. Here, 13% is not a parameter because we didn't get a chance to ask everyone. This 13% is an estimate of a parameter. Do you know what that's called?

That's right, a **statistic**!

We can define a statistic as a numerical measurement describing a characteristic of a sample of a population.

A statistic is just an estimation of a parameter. It is a number that attempts to describe an entire population by describing a subset of that population. This is necessary because you can never hope to give a survey to every single teenager or to every single smoker in the world. That's what the field of statistics is all about—taking samples of populations and running tests on these samples.

So, the next time you are given a statistic, just remember, that number only represents a sample of that population, not the entire pool of subjects.

How do we obtain and sample data?

If statistics is about taking samples of populations, it must be very important to know how we obtain these samples, and you'd be correct. Let's focus on just a few of the many ways of obtaining and sampling data.

Obtaining data

There are two main ways of collecting data for our analysis: **observational** and **experimentation**. Both these ways have their pros and cons, of course. They each produce different types of behavior and, therefore, warrant different types of analysis.

Observational

We might obtain data through *observational* means, which consists of measuring specific characteristics but not attempting to modify the subjects being studied. For example, you have a tracking software on your website that observes users' behavior on the website, such as length of time spent on certain pages and the rate of clicking on ads, all the while not affecting the user's experience, then that would be an observational study.

This is one of the most common ways to get data because it's just plain easy. All you have to do is observe and collect data. Observational studies are also limited in the types of data you may collect. This is because the observer (you) is not in control of the environment. You may only watch and collect natural behavior. If you are looking to induce a certain type of behavior, an observational study would not be useful.

Experimental

An **experiment** consists of a treatment and the observation of its effect on the subjects. Subjects in an experiment are called experimental units. This is usually how most scientific labs collect data. They will put people into two or more groups (usually just two) and call them the control and the experimental group.

The control group is exposed to a certain environment and then observed. The experimental group is then exposed to a different environment and then observed. The experimenter then aggregates data from both the groups and makes a decision about which environment was more favorable (favorable is a quality that the experimenter gets to decide).

In a marketing example, consider that we expose half of our users to a certain landing page with certain images and a certain style (website A), and we measure whether or not they sign up for the service. Then, we expose the other half to a different landing page, different images, and different styles (website B) and again measure whether or not they sign up. We can then decide which of the two sites performed better and should be used going further. This, specifically, is called an **A/B test**. Let's see an example in Python! Let's suppose we run the preceding test and obtain the following results as a list of lists:

```
results = [ ['A', 1], ['B', 1], ['A', 0], ['A', 0] … ]
```

Here, each object in the list result represents a subject (person). Each person then has the following two attributes:

- Which website they were exposed to, represented by a single character
- Whether or not they converted (0 for no and 1 for yes)

We can then aggregate and come up with the following results table:

```
users_exposed_to_A = []
users_exposed_to_B = []
# create two lists to hold the results of each individual website
```

Once we create these two lists that will eventually hold each individual conversion Boolean (0 or 1), we will iterate all of our results of the test and add them to the appropriate list, as shown:

```
for website, converted in results: # iterate through the results
  # will look something like website == 'A' and converted == 0
  if website == 'A':
    users_exposed_to_A.append(converted)
  elif website == 'B':
    users_exposed_to_B.append(converted)
```

Now, each list contains a series of 1s and 0s.

Remember that a 1 represents a user actually converting to the site after seeing that web page, and a 0 represents a user seeing the page and leaving before signing up/converting.

To get the total number of people exposed to website `A`, we can use the `len()` feature in Python, as illustrated:

```
len(users_exposed_to_A) == 188 #number of people exposed to website A
len(users_exposed_to_B) == 158 #number of people exposed to website B
```

To count the number of people who converted, we can use the `sum()` of the list, as shown:

```
sum(users_exposed_to_A) == 54 # people converted from website A
sum(users_exposed_to_B) == 48 # people converted from website B
```

If we subtract the length of the lists and the sum of the list, we are left with the number of people who did *not* convert for each site, as illustrated:

```
len(users_exposed_to_A) - sum(users_exposed_to_A) == 134 # did not
convert from website A

len(users_exposed_to_B) - sum(users_exposed_to_B) == 110 # did not
convert from website B
```

We can aggregate and summarize our results in the following table that represents our experiment of website conversion testing:

	Did not sign up	**Signed up**
Website A	134	54
Website B	110	48

The results of our A/B test

We can quickly drum up some descriptive statistics. We can say that the website conversion rates for the two websites are as follows:

- Conversion for website A: $\frac{54}{134+54} = .288$
- Conversion for website B: $\frac{48}{110+48} = .3$

Not much difference, but different nonetheless. Even though B has the higher conversion rate, can we really say that the version B significantly converts better? Not yet. To test the *statistical significance* of such a result, a hypothesis test should be used. These tests will be covered in depth in the next chapter, where we will revisit this exact same example and finish it using the proper statistical test.

Sampling data

Remember how statistics are the result of measuring a sample of a population. Well, we should talk about two very common ways to decide *who* gets the honor of being in the sample that we measure. We will discuss the main type of sampling, called random sampling, which is the most common way to decide our sample sizes and our sample members.

Probability sampling

Probability sampling is a way of sampling from a population, in which every person has a known probability of being chosen but that number *might* be a different probability than another user. The simplest (and probably the most common) probability sampling method is random sampling.

Random sampling

Suppose that we are running an A/B test and we need to figure out who will be in group A and who will be in group B. There are the following three suggestions from your data team:

- **Separate users based on location**: Users on the west coast are placed in group A, while users on the east coast are placed in group B
- **Separate users based on the time of day they visit the site**: Users who visit between 7 p.m. and 4 a.m. get site A, while the rest are placed in group B
- **Make it completely random**: Every new user has a 50/50 chance of being placed in either group

The first two are valid options for choosing samples and are fairly simple to implement, but they both have one fundamental flaw: they are both at risk of introducing a sampling bias.

A **sampling bias** occurs when the way the sample is obtained systemically favors some outcome over the target outcome.

It is not difficult to see why choosing option 1 or option 2 might introduce bias. If we chose our groups based on where they live or what time they log in, we are priming our experiment incorrectly and, now, we have much less control over the results.

Specifically, we are at risk of introducing a *confounding factor* into our analysis, which is bad news.

A **confounding factor** is a variable that we are not directly measuring but connects the variables that are being measured.

Basically, a confounding factor is like the missing element in our analysis that is invisible but affects our results.

In this case, option 1 is not taking into account the potential confounding factor of *geographical taste*. For example, if website A is unappealing, in general, to the west coast users, it will affect your results drastically.

Similarly, option 2 might introduce a temporal (time-based) confounding factor. What if website B is better viewed in a nighttime environment (which was reserved for A), and users are turned off to the style purely because of what time it is. These are both factors that we want to avoid, so, we should go with option 3, which is a random sample.

While sampling bias can cause confounding, it is a different concept than confounding. Options 1 and 2 were both sampling biases because we chose the samples incorrectly and were also examples of confounding factors because there was a third variable in each case that affected our decision.

A random sample is chosen such that every single member of a population has an equal chance of being chosen as any other member.

This is probably one of the easiest and most convenient ways to decide who will be a part of your sample. Everyone has the exact same chance of being in any particular group. Random sampling is an effective way of reducing the impact of confounding factors.

Unequal probability sampling

Recall that I previously said that a probability sampling might have different probabilities for different potential sample members. But what if this actually introduced problems? Suppose we are interested in measuring the happiness level of our employees. We already know that we can't ask every single person on the staff because that would be silly and exhausting. So, we need to take a sample. Our data team suggests random sampling and at first everyone high fives because they feel very smart and statistical. But then someone asks a seemingly harmless question—does anyone know the percentage of men/women who work here?

The high fives stop and the room goes silent.

This question is extremely important because sex is likely to be a confounding factor. The team looks into it and discovers a split of 75% men and 25% women in the company.

This means that if we introduce a random sample, our sample will likely have a similar split and, thus, favor the results for men and not women. To combat this, we can favor including more women than men in our survey in order to make the split of our sample less favored for men.

At first glance, introducing a favoring system in our random sampling seems like a bad idea, however, alleviating unequal sampling and, therefore, working to remove systematic bias among gender, race, disability, and so on is much more pertinent. A simple random sample, where everyone has the same chance as everyone else, is very likely to drown out the voices and opinions of minority population members. Therefore, it can be okay to introduce such a favoring system in your sampling techniques.

How do we measure statistics?

Once we have our sample, it's time to quantify our results. Suppose we wish to generalize the happiness of our employees or we want to figure out whether salaries in the company are very different from person to person.

These are some common ways of measuring our results.

Measures of center

Measures of center are how we define the middle, or center, of a dataset. We do this because sometimes we wish to make generalizations about data values. For example, perhaps we're curious about what the average rainfall in Seattle is or what the median height for European males is. It's a way to generalize a large set of data so that it's easier to convey to someone.

A measure of center is a value in the "middle" of a dataset.

However, this can mean different things to different people. Who's to say where the middle of a dataset is? There are so many different ways of defining the center of data. Let's take a look at a few.

The **arithmetic mean** of a dataset is found by adding up all of the values and then dividing it by the number of data values.

This is likely the most common way to define the center of data, but can be flawed! Suppose we wish to find the mean of the following numbers:

```
import numpy as np

np.mean([11, 15, 17, 14]) == 14.25
```

Simple enough, our average is `14.25` and all of our values are fairly close to it. But what if we introduce a new value: `31`?

```
np.mean([11, 15, 17, 14, 31]) == 17.6
```

This greatly affects the mean because the arithmetic mean is sensitive to outliers. The new value, `31`, is almost twice as large as the rest of the numbers and, therefore, *skews* the mean.

Another, and sometimes better, measure of center is the median.

The **median** is the number found in the middle of the dataset when it is sorted in order, as shown:

```
np.median([11, 15, 17, 14]) == 14.5
np.median([11, 15, 17, 14, 31]) == 15
```

Note how the introduction of `31` using the median did not affect the median of the dataset greatly. This is because the median is less sensitive to outliers.

When working with datasets with many outliers, it is sometimes more useful to use the median of the dataset, while if your data does not have many outliers and the data points are mostly close to one another, then the mean is likely a better option.

But how can we tell if the data is spread out? Well, we will have to introduce a new type of statistic.

Measures of variation

Measures of center are used to quantify the middle of the data, but now we will explore ways of measuring how "spread out" the data we collect is. This is a useful way to identify if our data has many outliers lurking inside. Let's start with an example.

Consider that we take a random sample of 24 of our friends on Facebook and wrote down how many friends that they had on Facebook. Here's the list:

```
friends = [109, 1017, 1127, 418, 625, 957, 89, 950, 946, 797, 981,
125, 455, 731, 1640, 485, 1309, 472, 1132, 1773, 906, 531, 742, 621]

np.mean(friends) == 789.1
```

The average of this list is just over 789. So, we could say that according to this sample, the average Facebook friend has 789 friends. But what about the person who only has 89 friends or the person who has over 1,600 friends? In fact, not a lot of these numbers are really that close to 789.

Well, how about we use the median, as shown, because the median generally is not as affected by outliers:

```
np.median(friends) == 769.5
```

The median is `769.5`, which is fairly close to the mean. Hmm, good thought, but still, it doesn't really account for how drastically different a lot of these data points are to one another. This is what statisticians call measuring the variation of data. Let's start by introducing the most basic measure of variation: the range. The range is simply the maximum value minus the minimum value, as illustrated:

```
np.max(friends) - np.min(friends) == 1684
```

The range tells us how far away the two most extreme values are. Now, typically, the range isn't widely used but it does have its use in application. Sometimes we wish to just know how spread apart the outliers are. This is most useful in scientific measurements or safety measurements.

Suppose a car company wants to measure how long it takes for an air bag to deploy. Knowing the average of that time is nice, but they also really want to know how spread apart the slowest time is versus the fastest time. This literally could be the difference between life and death.

Shifting back to the Facebook example, 1,684 is our range, but I'm not quite sure it's saying too much about our data. Now, let's take a look at the most commonly used measure of variation, the **standard deviation**.

I'm sure many of you have heard this term thrown around a lot and it might even incite a degree of fear, but what does it really mean? In essence, standard deviation, denoted by s when we are working with a sample of a population, measures how much data values deviate from the arithmetic mean.

It's basically a way to see how spread out the data is. There is a general formula to calculate the standard deviation, which is as follows:

$$s = \sqrt{\frac{\sum (x - \overline{x})^2}{n}}$$

Here:

- s is our sample standard deviation
- x is each individual data point.
- $\overline{x}$ is the mean of the data
- n is the number of data points

Before you freak out, let's break it down. For each value in the sample, we will take that value, subtract the arithmetic mean from it, square the difference, and, once we've added up every single point this way, we will divide the entire thing by *n*, the number of points in the sample. Finally, we take a square root of everything.

Without going into an in-depth analysis of the formula, think about it this way: it's basically derived from the distance formula. Essentially, what the standard deviation is calculating is a sort of average distance of how far the data values are from the arithmetic mean.

If you take a closer look at the formula, you will see that it actually makes sense:

- By taking $x-\overline{x}$, you are finding the literal difference between the value and the mean of the sample.
- By squaring the result, $(x-\overline{x})^2$, we are putting a greater penalty on outliers because squaring a large error only makes it much larger.
- By dividing by the number of items in the sample, we are taking (literally) the average squared distance between each point and the mean.
- By taking the square root of the answer, we are putting the number in terms that we can understand. For example, by squaring the number of friends minus the mean, we changed our units to friends square, which makes no sense. Taking the square root puts our units back to just "friends".

Let's go back to our Facebook example for a visualization and further explanation of this. Let's begin to calculate the standard deviation. So, we'll start calculating a few of them. Recall that the arithmetic mean of the data was just about 789, so, we'll use 789 as the mean.

We start by taking the difference between each data value and the mean, squaring it, adding them all up, dividing it by one less than the number of values, and then taking its square root. This would look as follows:

$$s = \sqrt{\frac{(109-789)^2 + (1017-789)^2 + \cdots + (621-789)^2}{24}}$$

On the other hand, we can take the Python approach and do all this programmatically (which is usually preferred).

```
np.std(friends) # == 425.2
```

What the number 425 represents is the spread of data. You could say that 425 is a kind of average distance the data values are from the mean. What this means, in simple words, is that this data is pretty spread out.

So, our standard deviation is about 425. This means that the number of friends that these people have on Facebook doesn't seem to be close to a single number and that's quite evident when we plot the data in a bar graph and also graph the mean as well as the visualizations of the standard deviation. In the following plot, every person will be represented by a single bar in the bar chart, and the height of the bars represent the number of friends that the individuals have:

```
import matplotlib.pyplot as plt
%matplotlib inline
y_pos = range(len(friends))

plt.bar(y_pos, friends)
plt.plot((0, 25), (789, 789), 'b-')
plt.plot((0, 25), (789+425, 789+425), 'g-')
plt.plot((0, 25), (789-425, 789-425), 'r-')
```

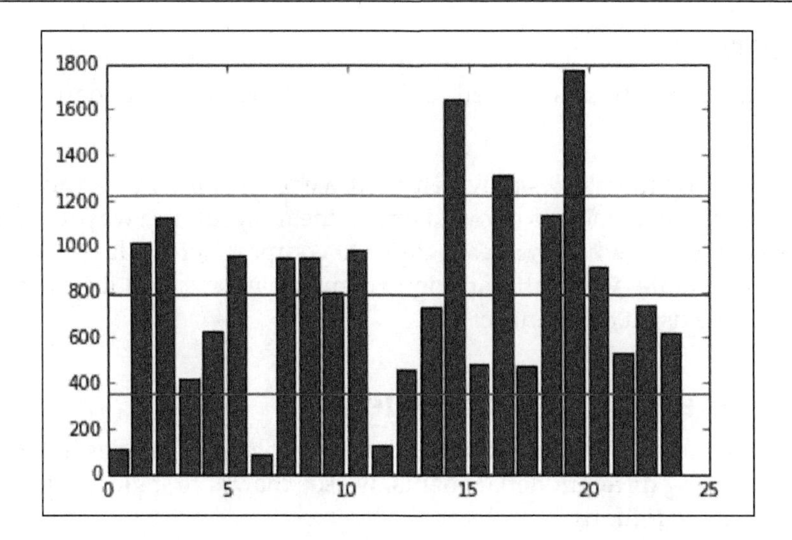

The blue line in the center is drawn at the mean (789), the red line on the bottom is drawn at the mean minus the standard deviation (789-425 = 364), and, finally, the green line towards the top is drawn at the mean plus the standard deviation (789+425 = 1,214).

Note how *most* of the data lives between the green and the red lines while the outliers live outside the lines. Namely, there are three people who have friend counts below the red line and three people who have a friend count above the green line.

It's important to mention that the units for standard deviation are, in fact, the same units as the data's units. So, in this example, we would say that the standard deviation is 425 friends on Facebook.

> Another measure of variation is the variance, as described in the previous chapter. The variance is simply the standard deviation, squared.

So, now we know that the standard deviation and variance is good for checking how spread out our data is, and that we can use it along with the mean to create a kind of range that a lot of our data lies in. But what if we want to compare the spread of two different datasets, maybe even with completely different units? That's where the coefficient of variation comes into play.

Definition

The **coefficient of variation** is defined as the ratio of the data's standard deviation to its mean.

This ratio (which, by the way, is only helpful if we're working in the ratio level of measurement, where division is allowed and is meaningful) is a way to standardize the standard deviation, which makes it easier to compare across datasets. We use this measure frequently when attempting to compare means, and it spreads across populations that exist at different scales.

Example – employee salaries

If we look at the mean and standard deviation of employees' salaries in the same company but among different departments, we see that, at first glance, it may be tough to compare variations.

Salaries of Company XYZ

Department	Mean Salary	SD	CoV
Mailroom	$25,000	$2,000	8.0%
Human Resources	$52,000	$7,000	13.5%
Executive	$124,000	$42,000	33.9%

This is especially true when the mean salary of one department is $25,000, while another department has a mean salary in the six-figure area.

However, if we look at the last column, which is our coefficient of variation, it becomes clearer that the people in the executive department may be getting paid more but employees in the executive department are getting wildly different salaries. This is probably because the CEO is earning way more than an office manager, who is still in the executive department, which makes the data very spread out.

On the other hand, everyone in the mailroom, while not making as much money, are making just about the same as everyone else in the mailroom, which is why their coefficient of variation is only 8%.

With measures of variation, we can begin to answer big questions, such as how spread out this data is or how we can come up with a good range that most of the data falls in.

Measures of relative standing

We can combine both the measures of centers and variations to create measures of relative standings.

Measures of variation measure where particular data values are positioned, relative to the entire dataset.

Let's begin by learning a very important value in statistics, the z-score.

The **z-score** is a way of telling us how far away a single data value is from the mean.

The z-score of a x data value is as follows:

$$z = \frac{x - \overline{x}}{s}$$

Where:

- x is the data point
- $\overline{x}$ is the mean
- s is the standard deviation.

Remember that the standard deviation was (sort of) an average distance that the data is from the mean, and, now, the z-score is an individualized value for each particular data point. We can find the z-score of a data value by subtracting it from the mean and dividing it by the standard deviation. The output will be the standardized distance a value is from a mean. We use the z-score all over statistics. It is a very effective way of normalizing data that exists on very different scales, and also to put data in context of their mean.

Let's take our previous data on the number of friends on Facebook and standardize the data to the z-score. For each data point, we will find its z-score by applying the preceding formula. We will take each individual, subtract the average friends from the value, and divide that by the standard deviation, as shown:

```
z_scores = []

m = np.mean(friends)  # average friends on Facebook
s = np.std(friends)   # standard deviation friends on Facebook

for friend in friends:
```

```
    z = (friend - m)/s  # z-score
    z_scores.append(z)  # make a list of the scores for plotting
```

Now, let's plot these z-scores on a bar chart. The following chart shows the same individuals from our previous example using friends on Facebook, but, instead of the bar height revealing the raw number of friends, now each bar is the z-score of the number of friends they have on Facebook. If we graph the z-scores, we'll notice a few things:

```
plt.bar(y_pos, z_scores)
```

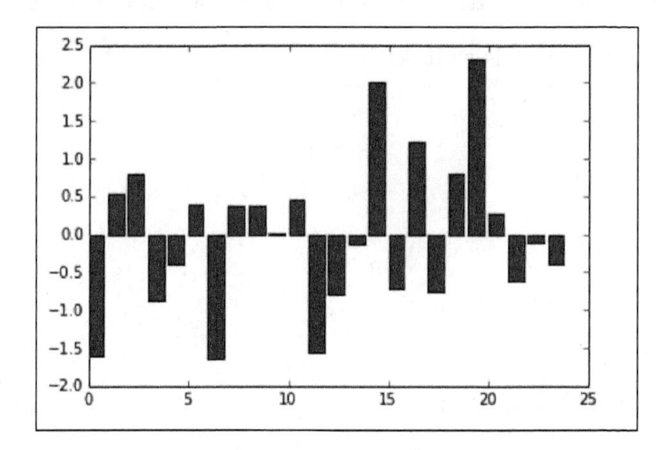

- We have negative values (meaning that the data point is below the mean)
- The bars' lengths no longer represent the raw number of friends, but the degree to which that friend count differs from the mean

This chart makes it very easy to pick out the individuals with much lower and higher friends on an average. For example, the individual at index 0 has fewer friends on an average (they had 109 friends where the average was 789).

What if we want to graph the standard deviations? Recall that we earlier graphed three horizontal lines: one at the mean, one at the mean plus the standard deviation ($x+s$), and one at the mean minus the standard deviation ($x-s$).

If we plug in these values into the formula for the z-score, we will get:

Z-score of ($\bar{x}$x) $= \frac{x-\bar{x}}{s} = \frac{0}{s} = 0$

$$\text{Z-score of } (x+s) = \frac{(\bar{x}+s)-\bar{x}}{s} = \frac{s}{s} = 1$$

$$\text{Z-score of } (x-s)\ \frac{(\bar{x}-s)-\bar{x}}{s} = \frac{-s}{s} = -1$$

This is no coincidence! When we standardize the data using the z-score, our standard deviations become the metric of choice. Let's see our new graph with the standard deviations plotted:

```
plt.bar(y_pos, z_scores)
plt.plot((0, 25), (1, 1), 'g-')
plt.plot((0, 25), (0, 0), 'b-')
plt.plot((0, 25), (-1, -1), 'r-')
```

The preceding code is adding in the following three lines:

- A blue line at *y = 0* that represents zero standard deviations away from the mean (which is on the *x* axis)
- A green line that represents one standard deviation above the mean
- A red line that represents one standard deviation below the mean

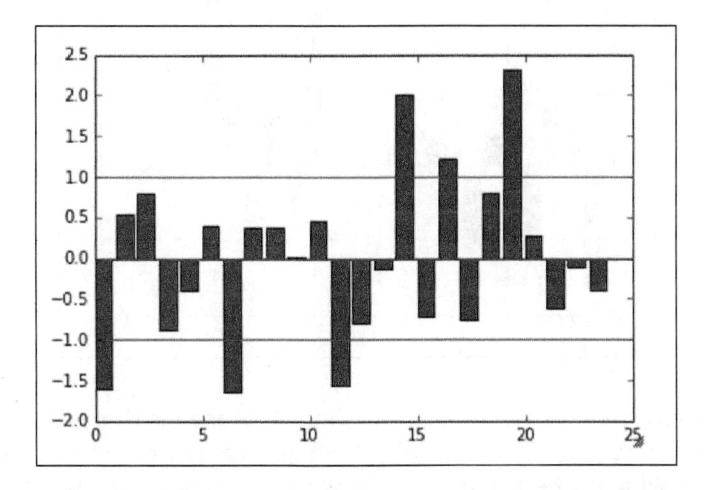

The colors of the lines match up with the lines drawn in the earlier graph of the raw friend count. If you look carefully, the same people still fall outside of the green and the red lines. Namely, the same three people still fall below the red (lower) line, and the same three people fall above the green (upper) line.

Under this scaling, we can also use statements as follows:

- This data point is over one standard deviation away from the mean:

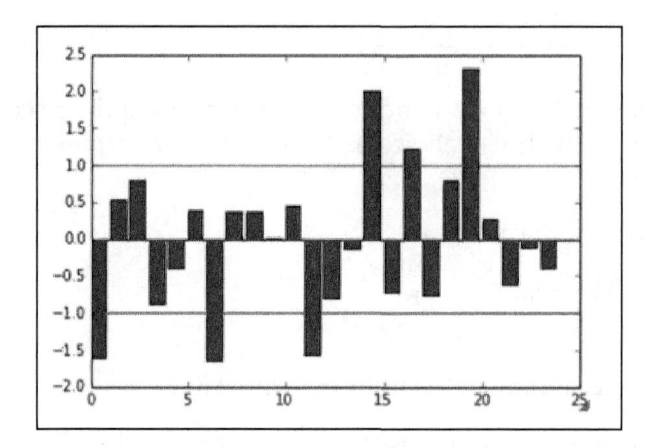

- This person has a friend count within one standard deviation from the mean:

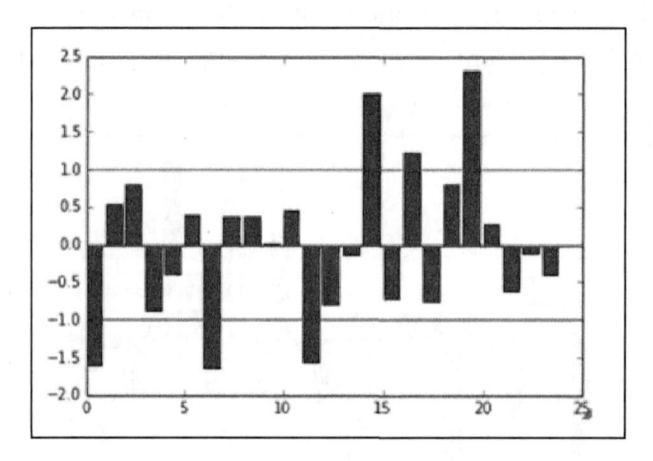

Z-scores are an effective way to *standardize* data. This means that we can put the entire set on the same scale. For example, if we also measure each person's general happiness scale (which is between 0 and 1), we might have a dataset similar to the following dataset:

```
friends = [109, 1017, 1127, 418, 625, 957, 89, 950, 946, 797, 981,
125, 455, 731, 1640, 485, 1309, 472, 1132, 1773, 906, 531, 742, 621]

happiness = [.8, .6, .3, .6, .6, .4, .8, .5, .4, .3, .3, .6, .2, .8,
1, .6, .2, .7, .5, .3, .1, 0, .3, 1]
```

```
import pandas as pd

df = pd.DataFrame({'friends':friends, 'happiness':happiness})
df.head()
```

	friends	happiness
0	109	0.8
1	1017	0.6
2	1127	0.3
3	418	0.6
4	625	0.6

These data points are on two different dimensions, each with a very different scale. The friend count can be in the thousands while our happiness score is stuck between 0 and 1.

To remedy this (and for some statistical/machine learning modeling, this concept will become essential), we can simply standardize the dataset using a prebuilt standardization package in `scikit-learn`, as follows:

```
from sklearn import preprocessing

df_scaled = pd.DataFrame(preprocessing.scale(df), columns = ['friends_
scaled', 'happiness_scaled'])

df_scaled.head()
```

This code will scale both the friends and happiness columns simultaneously, thus revealing the z-score for each column. It is important to note that by doing this, the preprocessing module in `sklearn` is doing the following things separately for each column:

- Finding the mean of the column
- Finding the standard deviation of the column
- Applying the z-score function to each element in the column

The result is two columns, as shown, that exist on the same scale as each other even if they were not previously:

	friends_scaled	happiness_scaled
0	-1.599495	1.153223
1	0.536040	0.394939
2	0.794750	-0.742486
3	-0.872755	0.394939
4	-0.385909	0.394939

Now, we can plot friends and happiness on the same scale and the graph will at least be readable.

```
df_scaled.plot(kind='scatter', x = 'friends_scaled', y = 'happiness_
scaled')
```

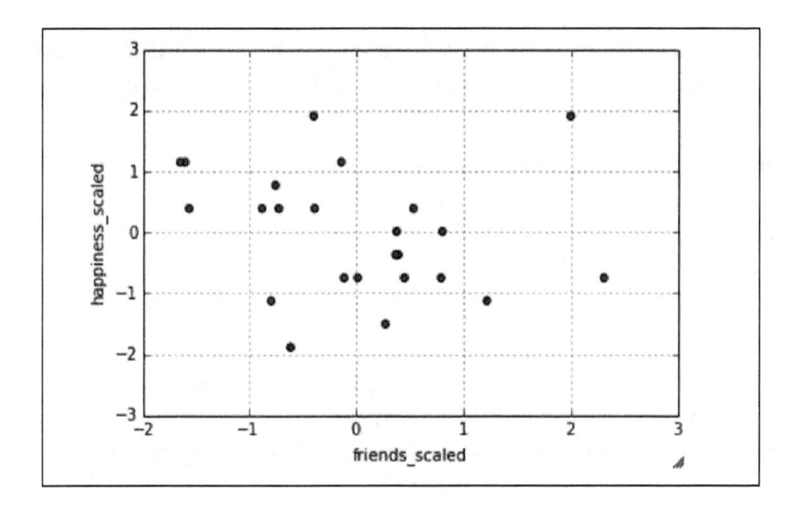

Now our data is standardized to the z-score and this scatter plot is fairly easily interpretable! In later chapters, this idea of standardization will not only make our data more interpretable, but it will also be essential in our model optimization. Many machine learning algorithms will require us to have standardized columns as they are reliant on the notion of scale.

The insightful part – correlations in data

Throughout this book, we will discuss the difference between having data and having actionable insights about your data. Having data is only one step to a successful data science operation. Being able to obtain, clean, and plot data helps to tell the story that the data has to offer but cannot reveal the moral. In order to take this entire example one step further, we will look at the relationship between having friends on Facebook and happiness.

In subsequent chapters, we will look at a specific machine learning algorithm that attempts to find relationships between quantitative features, called linear regression, but we do not have to wait until then to begin to form hypotheses. We have a sample of people, a measure of their online social presence and their reported happiness. The question of the day here is—can we find a relationship between the number of friends on Facebook and overall happiness?

Now, obviously, this is a big question and should be treated respectfully. Experiments to answer this question should be conducted in a laboratory setting, but we can begin to form a hypothesis about this question. Given the nature of our data, we really only have the following three options for a hypothesis:

- There is a positive association between the number of online friends and happiness (as one goes up, so does the other)
- There is a negative association between them (as the number of friends goes up, your happiness goes down)
- There is no association between the variables (as one changes, the other doesn't really change that much)

Can we use basic statistics to form a hypothesis about this question? I say we can! But first, we must introduce a concept called correlation.

Correlation coefficients are a quantitative measure that describe the strength of association/relationship between two variables.

The correlation between two sets of data tells us about how they move together. Would changing one help us predict the other? This concept is not only interesting in this case, but it is one of the core assumptions that many machine learning models make on data. For many prediction algorithms to work, they rely on the fact that there is some sort of relationship between the variables that we are looking at. The learning algorithms then exploit this relationship in order to make accurate predictions.

A few things to note about a correlation coefficient are as follows:

- It will lie between -1 and 1
- The greater the absolute value (closer to -1 or 1), the stronger the relationship between the variables:
    - The strongest correlation is a -1 or a 1
    - The weakest correlation is a 0
- A positive correlation means that as one variable increases, the other one tends to increase as well
- A negative correlation means that as one variable increases, the other one tends to decrease

We can use Pandas to quickly show us correlation coefficients between every feature and every other feature in the Dataframe, as illustrated:

```
# correlation between variables
df.corr()
```

	friends	**happiness**
friends	1.000000	-0.216199
happiness	-0.216199	1.000000

This table shows the correlation between friends and happiness. Note the first two things, shown as follows:

- The diagonal of the matrix is filled with positive is. This is because they represent the correlation between the variable and itself, which, of course, forms a perfect line, making the correlation perfectly positive!
- The matrix is symmetrical across the diagonal. This is true for any correlation matrix made in Pandas.

There are a few caveats to trusting the correlation coefficient. One is that, in general, a correlation will attempt to measure a *linear* relationship between variables. This means that if there is no visible correlation revealed by this measure, it does not mean that there is no relationship between the variables, only that there is no line of best fit that goes through the lines easily. There might be a *non-linear* relationship that defines the two variables.

It is important to realize that causation is not implied by correlation. Just because there is a weak negative correlation between these two variables does not necessarily mean that your overall happiness decreases as the number of friends you keep on Facebook goes up. This causation must be tested further and, in later chapters, we will attempt to do just that.

To sum up, we can use correlation to make hypotheses about the relationship between variables, but we will need to use more sophisticated statistical methods and machine learning algorithms to solidify these assumptions and hypotheses.

The Empirical rule

Recall that a normal distribution is defined as having a specific probability distribution that resembles a bell curve. In statistics, we love it when our data behaves *normally*. For example, if we have data that resembles a normal distribution, like so:

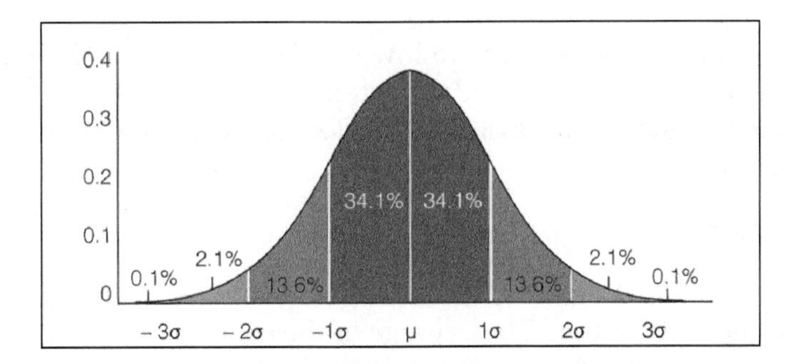

The **Empirical rule** states that we can expect a certain amount of data to live between sets of standard deviations. Specifically, the Empirical rule states for data that is distributed normally:

- about 68% of the data fall within **1** standard deviation
- about 95% of the data fall within **2** standard deviations
- about 99.7% of the data fall within **3** standard deviations

For example, let's see if our Facebook friends' data holds up to this. Let's use our Dataframe to find the percentage of people that fall within 1, 2, and 3 standard deviations of the mean, as shown:

```
# finding the percentage of people within one standard deviation of
the mean
within_1_std = df_scaled[(df_scaled['friends_scaled'] <= 1) & (df_
scaled['friends_scaled'] >= -1)].shape[0]
```

```
within_1_std / float(df_scaled.shape[0])
# 0.75


# finding the percentage of people within two standard deviations of
the mean
within_2_std = df_scaled[(df_scaled['friends_scaled'] <= 2) & (df_
scaled['friends_scaled'] >= -2)].shape[0]
within_2_std / float(df_scaled.shape[0])
# 0.916


# finding the percentage of people within three standard deviations of
the mean
within_3_std = df_scaled[(df_scaled['friends_scaled'] <= 3) & (df_
scaled['friends_scaled'] >= -3)].shape[0]
within_3_std / float(df_scaled.shape[0])
# 1.0
```

We can see that our data does seem to follow the Empirical rule. About 75% of the people are within a single standard deviation of the mean. About 92% of the people are within two standard deviations, and all of them are within three standard deviations.

Example – exam scores

Let's say that we're measuring the scores of an exam and the scores generally have a bell-shaped normal distribution. The average of the exam was 84% and the standard deviation was 6%. We can say, with approximate certainty, that:

- About 68% of the class scored between 78% and 90% because 78 is 6 units below 84, and 90 is 6 units above 84
- If we were asked what percentage of the class scored between 72 and 96%, we would notice that 72 is 2 standard deviations below the mean, and 96 is 2 standard deviations above the mean, so, the Empirical rule tells us that about 95% of the class scored in that range.

However, not all data is normally distributed, so, we can't always use the Empirical rule. We have another theorem that helps us analyze any kind of distribution. In the next chapter, we will go in depth about when we can assume the normal distribution. This is because many statistical tests and hypotheses require the underlying data to come from a normally distributed population.

Previously, when we standardized our data to the z-score, we did not require the normal distribution assumption.

Summary

In this chapter, we covered much of the basic statistics required by most data scientists. Everything from how we obtain/sample data to how to standardize data according to the z-score and applications of the Empirical rule was covered.

In the next chapter, we will look at much more advanced applications of statistics. One thing that we will consider is how to use hypothesis tests on data that we can assume to be normal. As we use these tests, we will also quantify our errors and identify the best practices to solve these errors.

8
Advanced Statistics

We are concerned with making inferences about entire populations based on certain samples of data. We will be using hypothesis tests along with different estimation tests in order to gain a better understanding of populations, given samples of data.

The key topics that we will cover in this chapter are as follows:

- Point estimates
- Confidence intervals
- The central limit theorem
- Hypothesis testing

Point estimates

Recall that, in the previous chapter, we mentioned how difficult it was to obtain a population parameter; so, we had to use sample data to calculate a statistic that was an estimate of a parameter. When we make these estimates, we call them point estimates.

A **point estimate** is an estimate of a population parameter based on sample data.

We use point estimates to estimate population means, variances, and other statistics. To obtain these estimates, we simply apply the function that we wish to measure for our population to a sample of the data. For example, suppose there is a company of 9,000 employees and we are interested in ascertaining the average length of breaks taken by employees in a single day. As we probably cannot ask every single person, we will take a sample of the 9,000 people and take a mean of the sample. This sample mean will be our point estimate.

The following code is broken into three parts:

- We will use the probability distribution, known as the **Poisson distribution**, to randomly generate 9,000 answers to the question: *for how many minutes in a day do you usually take breaks?* This will represent our "population". Remember, from *Chapter 6, Advanced Probability*, that the **Poisson random variable** is used when we know the average value of an event and wish to model a distribution around it.

Note that this average value is not usually known. I am calculating it to show the difference between our parameter and our statistic. I also set a random seed in order to encourage reproducibility (this allows us to get the same *random* numbers each time.)

- We will take a sample of 100 employees (using the Python random sample method) and find a point estimate of a mean (called a sample mean).

Note that this is just over 1% of our population.

- Compare our sample mean (the mean of the sample of 100 employees) to our population mean.

Let's take a look at the following code:

```
np.random.seed(1234)

long_breaks = stats.poisson.rvs(loc=10, mu=60, size=3000)
# represents 3000 people who take about a 60 minute break
```

The `long_breaks` variable represents `3000` answers to the question: *how many minutes on an average do you take breaks for?*, and these answers will be on the longer side. Let's see a visualization of this distribution, shown as follows:

```
pd.Series(long_breaks).hist()
```

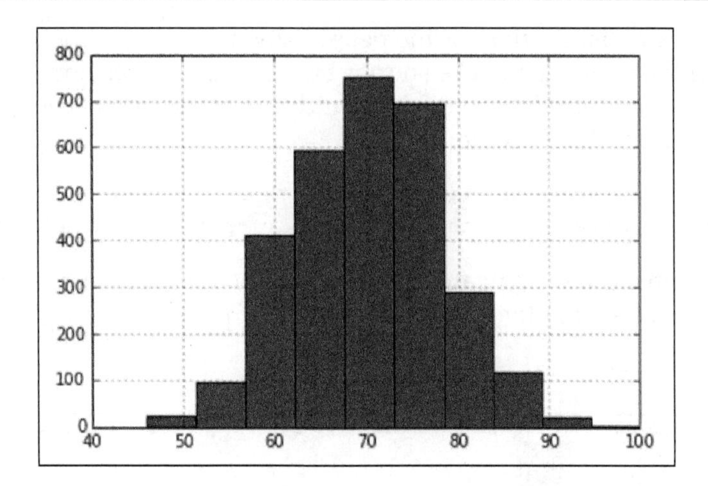

We see that our average of `60` minutes is to the left of the distribution. Also, because we only sampled `3000` people, our bins are at their highest around **700-800** people.

Now, let's model `6000` people who take, on an average, about `15` minutes' worth of breaks. Let's again use the Poisson distribution to simulate `6000` people, as shown:

```
short_breaks = stats.poisson.rvs(loc=10, mu=15, size=6000)
# represents 6000 people who take about a 15 minute break
pd.Series(short_breaks).hist()
```

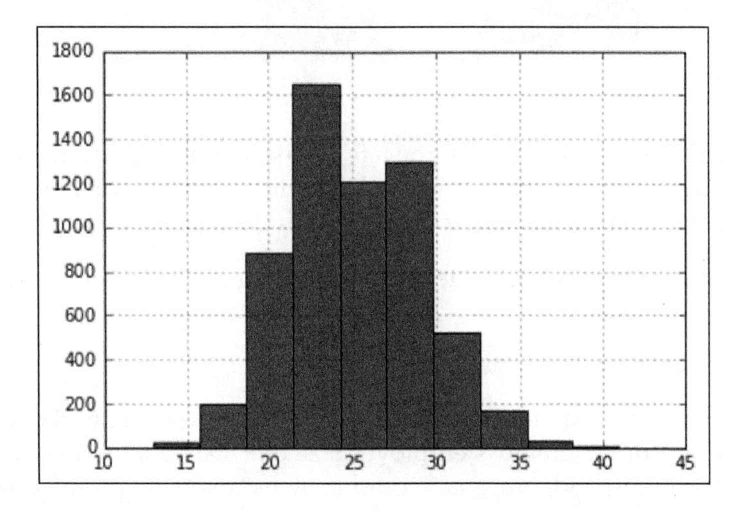

Okay, so, we have a distribution for the people who take longer breaks and a distribution for the people who take shorter breaks. Again, note how our average break length of `15` minutes falls to the left-hand side of the distribution, and note that the tallest bar is about **1600** people.

```
breaks = np.concatenate((long_breaks, short_breaks))
# put the two arrays together to get our "population" of 9000 people
```

The `breaks` variable is the amalgamation of all the `9000` employees, both long and short break takers. Let's see the entire distribution of people in a single visualization:

```
pd.Series(breaks).hist()
```

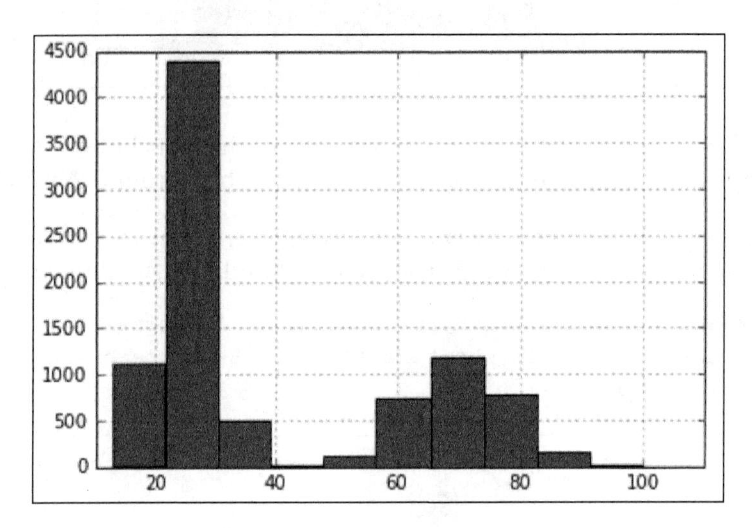

We see how we have two humps. On the left, we have our larger hump of people who take about a 15 minute break, and on the right, we have a smaller hump of people who take longer breaks. Later on, we will investigate this graph further.

We can find the total average break length by running the following code:

```
breaks.mean()
# 39.99 minutes is our parameter.
```

Our average company break length is about 40 minutes. Remember that our population is the entire company's employee size of 9,000 people, and *our parameter is 40 minutes*. In the real world, our goal would be to estimate the population parameter because we would not have the resources to ask every single employee in a survey their average break length for many reasons. Instead, we will use a point estimate.

So, to make our point, we want to simulate a world where we ask 100 random people about the length of their breaks. To do this, let's take a random sample of `100` employees out of the 9,000 employees we simulated, as shown:

```
sample_breaks = np.random.choice(a = breaks, size=100)
# taking a sample of 100 employees
```

Now, let's take the mean of the sample and subtract it from the population mean and see how far off we were:

```
breaks.mean() - sample_breaks.mean()
# difference between means is 4.09 minutes, not bad!
```

This is extremely interesting, because with only about 1% of our population (100 out of 9,000), we were able to get within 4 minutes of our population parameter and get a very accurate estimate of our population mean. Not bad!

Here, we calculated a point estimate for the mean, but we can also do this for proportion parameters. By proportion, I am referring to a ratio of two quantitative values.

Let's suppose that in a company of 10,000 people, our employees are 20% white, 10% black, 10% Hispanic, 30% Asian, and 30% identify as other. We will take a sample of 1,000 employees and see if their race proportions are similar.

```
employee_races = (["white"]*2000) + (["black"]*1000) +\
                 (["hispanic"]*1000) + (["asian"]*3000) +\
                 (["other"]*3000)
```

`employee_races` represents our employee population. For example, in our company of 10,000 people, 2,000 people are white (20%) and 3,000 people are Asian (30%).

Let's take a random sample of 1,000 people, as shown:

```
demo_sample = random.sample(employee_races, 1000)    # Sample 1000
values

for race in set(demo_sample):
    print( race + " proportion estimate:" )
    print( demo_sample.count(race)/1000. )
```

The output obtained would be as follows:

```
hispanic proportion estimate:
0.103
white proportion estimate:
0.192
```

```
other proportion estimate:
0.288
black proportion estimate:
0.1
asian proportion estimate:
0.317
```

We can see that the race proportion estimates are very close to the underlying population's proportions. For example, we got 10.3% for Hispanic in our sample and the population proportion for Hispanic was 10%.

Sampling distributions

In *Chapter 7, Basic Statistics*, we mentioned how much we love when data follows the normal distribution. One of the reasons for this is that many statistical tests (including the ones we will use in this chapter) rely on data that follows a normal pattern, and for the most part, a lot of real-world data is not normal (surprised?). Take our employee break data for example, you might think I was just being fancy creating data using the Poisson distribution, but I had a reason for this—I specifically wanted non-normal data, as shown:

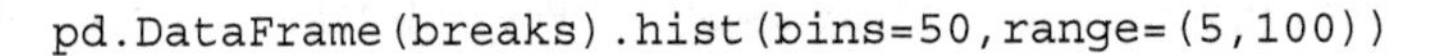

```
pd.DataFrame(breaks).hist(bins=50,range=(5,100))
```

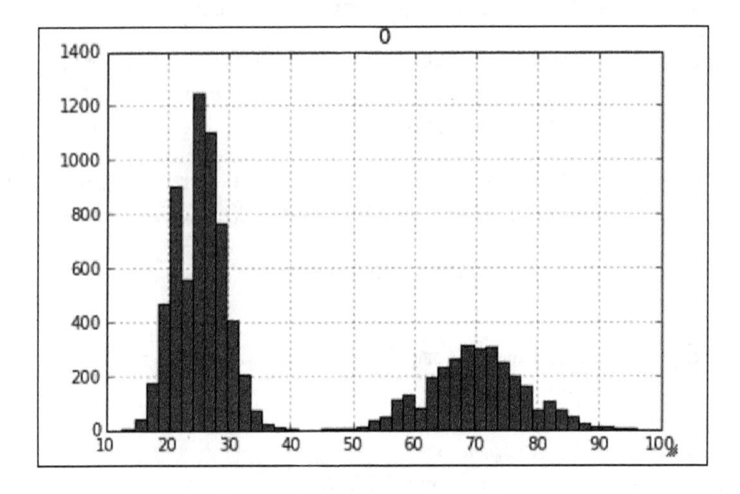

As you can see, our data is definitely not following a normal distribution, it appears to be **bi-modal**, which means that there are two peaks of break times, at around **25** and **70** minutes. As our data is not normal, many of the most popular statistics tests may not apply, however, if we follow the given procedure, we can create normal data! Think I'm crazy? Well, see for yourself.

First off, we will need to utilize what is known as a **sampling distribution**, which is a distribution of point estimates of several samples of the same size. Our procedure for creating a sampling distribution will be the following:

1. Take 500 different samples of the break times of size 100 each.
2. Take a histogram of these 500 different point estimates (revealing their distribution).

The number of elements in the sample (100) was arbitrary, but large enough to be a representative sample of the population. The number of samples I took (500) was also arbitrary, but large enough to ensure that our data would converge to a normal distribution:

```
point_estimates = []

for x in range(500):          # Generate 500 samples
    sample = np.random.choice(a= breaks, size=100)
#take a sample of 100 points

point_estimates.append( sample.mean() )
# add the sample mean to our list of point estimates

pd.DataFrame(point_estimates).hist()
# look at the distribution of our sample means
```

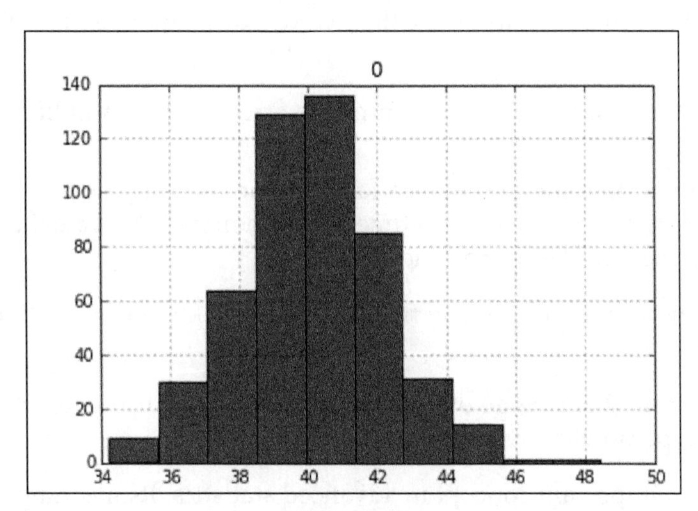

Behold! The sampling distribution of the sample mean appears to be normal even though we took data from an underlying bimodal population distribution. It is important to note that the bars in this histogram represent the average break length of 500 samples of employees, where each sample has 100 people in it. In other words, a sampling distribution is a distribution of several point estimates.

Our data converged to a normal distribution because of something called the **central limit theorem**, which states that the sampling distribution (the distribution of point estimates) will approach a normal distribution as we increase the number of samples taken.

What's more, as we take more and more samples, the mean of the sampling distribution will approach the true population mean, as shown:

```
breaks.mean() - np.array(point_estimates).mean()
# .047 minutes difference
```

This is actually a very exciting result because it means that we can get even closer than a single point estimate by taking multiple point estimates and utilizing the central limit theorem!

In general, as we increase the number of samples taken, our estimate will get closer to the parameter (actual value).

Confidence intervals

While point estimates are okay estimates of a population parameter and sampling distributions are even better, there are the following two main issues with these approaches:

- Single point estimates are very prone to error (due to sampling bias among other things)
- Taking multiple samples of a certain size for sampling distributions might not be feasible, and may sometimes be even more infeasible than actually finding the population parameter

For these reasons and more, we may turn to a concept, known as confidence interval, to find statistics.

A **confidence interval** is a range of values based on a point estimate that contains the true population parameter at some confidence level.

Confidence is an important concept in advanced statistics. Its meaning is sometimes misconstrued. Informally, a confidence level does *not* represent a "probability of being correct"; instead, it represents the frequency that the obtained answer will be accurate. For example, if you want to have a 95% chance of capturing the true population parameter using only a single point estimate, we would have to set our confidence level to 95%.

[Higher confidence levels result in wider (larger) confidence intervals in order to be *more sure*.]

Calculating a confidence interval involves finding a point estimate, and then, incorporating a margin of error to create a range. The **margin of error** is a value that represents our certainty that our point estimate is accurate and is based on our desired confidence level, the variance of the data, and how big your sample is. There are many ways to calculate confidence intervals; for the purpose of brevity and simplicity, we will look at a single way of taking the confidence interval of a population mean. For this confidence interval, we need the following:

- A point estimate. For this, we will take our sample mean of break lengths from our previous example.
- An estimate of the population standard deviation, which represents the variance in the data.
    - This is calculated by taking the sample standard deviation (the standard deviation of the sample data) and dividing that number by the square root of the population size.
- The degrees of freedom (which is the -1 sample size).

Obtaining these numbers might seem arbitrary but, trust me, there is a reason for all of them. However, again for simplicity, I will use prebuilt Python modules, as shown, to calculate our confidence interval and, then, demonstrate its value:

```
sample_size = 100
# the size of the sample we wish to take

sample = np.random.choice(a= breaks, size = sample_size)
# a sample of sample_size taken from the 9,000 breaks population from
before

sample_mean = sample.mean()
# the sample mean of the break lengths sample

sample_stdev = sample.std()
# sample standard deviation

sigma = sample_stdev/math.sqrt(sample_size)
# population standard deviation estimate

stats.t.interval(alpha = 0.95,              # Confidence level 95%
                 df= sample_size - 1,       # Degrees of freedom
```

```
                    loc = sample_mean,          # Sample mean
                    scale = sigma)              # Standard deviation
# (36.36, 45.44)
```

To reiterate, this range of values (from `36.36` to `45.44`) represents a confidence interval for the average break time with a 95% confidence.

We already know that our population parameter is `39,99`, and note that the interval includes the population mean of `39.99`.

I mentioned earlier that the confidence level was not a percentage of accuracy of our interval but the percent chance that the interval would even contain the population parameter at all.

To better understand the confidence level, let's take 10,000 confidence intervals and see how often our population mean falls in the interval. First, let's make a function, as illustrated, that makes a single confidence interval from our breaks data:

```
# function to make confidence interval
def makeConfidenceInterval():
    sample_size = 100
    sample = np.random.choice(a= breaks, size = sample_size)

    sample_mean = sample.mean()
    # sample mean

    sample_stdev = sample.std()
    # sample standard deviation

    sigma = sample_stdev/math.sqrt(sample_size)
    # population Standard deviation estimate

    return stats.t.interval(alpha = 0.95, df= sample_size - 1, loc =
sample_mean, scale = sigma)
```

Now that we have a function that will create a single confidence interval, let's create a procedure that will test the probability that a single confidence interval will contain the true population parameter, `39.99`:

1. Take 10,000 confidence intervals of the sample mean.
2. Count the number of times that the population parameter falls into our confidence intervals.

3. Output the ratio of the number of times the parameter fell into the interval by 10,000:

```
times_in_interval = 0.
for i in range(10000):
    interval = makeConfidenceInterval()
    if 39.99 >= interval[0] and 39.99 <= interval[1]:
    # if 39.99 falls in the interval
        times_in_interval += 1

print times_in_interval / 10000
# 0.9455
```

Success! We see that about 95% of our confidence intervals contained our actual population mean. Estimating population parameters through point estimates and confidence intervals is a relatively simple and powerful form of statistical inference.

Let's also take a quick look at how the size of confidence intervals changes as we change our confidence level. Let's calculate confidence intervals for multiple confidence levels and look at how large the intervals are by looking at the difference between the two numbers. Our hypothesis will be that as we make our confidence level larger, we will likely see larger confidence intervals to be surer that we catch the true population parameter:

```
for confidence in (.5, .8, .85, .9, .95, .99):
    confidence_interval = stats.t.interval(alpha = confidence, df=
sample_size - 1, loc = sample_mean, scale = sigma)

    length_of_interval = round(confidence_interval[1] - confidence_
interval[0], 2)
    # the length of the confidence interval

    print "confidence {0} has a interval of size {1}".
format(confidence, length_of_interval)

confidence 0.5 has an interval of size 2.56
confidence 0.8 has an interval of size 4.88
confidence 0.85 has an interval of size 5.49
confidence 0.9 has an interval of size 6.29
confidence 0.95 has an interval of size 7.51
confidence 0.99 has an interval of size 9.94
```

We can see that as we wish to be "more confident" in our interval, our interval expands in order to compensate.

Next, we will take our concept of confidence levels and look at statistical hypothesis testing in order to both expand on these topics and also create (usually) even more powerful statistical inferences.

Hypothesis tests

Hypothesis tests are one of the most widely used tests in statistics. They come in many forms; however, all of them have the same basic purpose.

A **hypothesis test** is a statistical test that is used to ascertain whether we are allowed to assume that a certain condition is true for the entire population, given a data sample. Basically, a hypothesis test is a test for a certain hypothesis that we have about an entire population. The result of the test then tells us whether we should believe the hypothesis or reject it for an alternative one.

You can think of the hypothesis tests' framework to determine whether the observed sample data deviates from what was to be expected from the population itself. Now this sounds like a difficult task but, luckily, Python comes to the rescue and includes built-in libraries to conduct these tests easily.

A hypothesis test generally looks at two opposing hypotheses about a population. We call them the **null hypothesis** and the **alternative hypothesis**. The null hypothesis is the statement being tested and is the *default correct* answer; it is our starting point and our original hypothesis. The alternative hypothesis is the statement that opposes the null hypothesis. Our test will tell us which hypothesis we should trust and which we should reject.

Based on sample data from a population, a hypothesis test determines whether or not to *reject* the null hypothesis. We usually use a **p-value** (which is based on our significance level) to make this conclusion.

A very common misconception is that statistical hypothesis tests are designed to select the more *likely* of the two hypotheses. This is incorrect. A hypothesis test will default to the null hypothesis *until* there is enough data to support the alternative hypothesis.

The following are some examples of questions you can answer with a hypothesis test:

- Does the mean break time of employees differ from 40 minutes?
- Is there a difference between people who interacted with website A and people who interacted with website B (A/B testing)?

- Does a sample of coffee beans vary significantly in taste from the entire population of beans?

Conducting a hypothesis test

There are multiple types of hypothesis tests out there, and among them are dozens of different procedures and metrics. Nonetheless, there are five basic steps that most hypothesis tests follow, which are as follows:

1. Specify the hypotheses:
   - Here, we formulate our two hypotheses: the null and the alternative
   - We usually use the notation of H_0 to represent the null hypothesis and H_a to represent our alternative hypothesis
2. Determine the sample size for the test sample:
   - This calculation depends on the chosen test. Usually, we have to determine a proper sample size in order to utilize theorems, such as the central limit theorem, and assume the normality of data.
3. Choose a significance level (usually called alpha or α):
   - A significance level of 0.05 is common
4. Collect the data:
   - They collect a sample of data to conduct the test
5. Decide whether to reject or fail to reject the null hypothesis:
   - This step changes slightly based on the type of test being used. The final result will either yield in rejecting the null hypothesis in favor of the alternative or failing to reject the null hypothesis.

In this chapter, we will look at the following three types of hypothesis tests:

- One-sample t-tests
- Chi-square goodness of fit
- Chi-square test for association/independence

There are many more tests. However, these three are a great combination of distinct, simple, and powerful tests. One of the biggest things to consider when choosing which test we should implement is the type of data we are working with, specifically, are we dealing with continuous or categorical data. In order to truly see the effects of a hypothesis, I suggest we dive right into an example. First, let's look at the use of a t-tests to deal with continuous data.

One sample t-tests

The **one sample t-test** is a statistical test used to determine whether a quantitative (numerical) data sample differs *significantly* from another dataset (the population or another sample). Suppose, in our previous employee break time example, we look, specifically, at the engineering department's break times, as shown:

```
long_breaks_in_engineering = stats.poisson.rvs(loc=10, mu=55,
size=100)

short_breaks_in_engineering = stats.poisson.rvs(loc=10, mu=15,
size=300)

engineering_breaks = np.concatenate((long_breaks_in_engineering,
short_breaks_in_engineering))

print breaks.mean()
# 39.99

print engineering_breaks.mean()
# 34.825
```

Note that I took the same approach as making the original break times, but with the following two differences:

- I took a smaller sample from the Poisson distribution (to simulate that we took a sample of 400 people from the engineering department)
- Instead of using a `mu` of `60` as before, I used `55` to simulate the fact that the engineering department's break behavior isn't exactly the same as the company's behavior as a whole

It is easy to see that there seems to be a difference (of over 5 minutes) between the engineering department and the company as a whole. We usually don't have the entire population and the population parameters at our disposal, but I have them simulated in order to see the example work. So, even though we (the omniscient readers) can see a difference, we will assume that we know nothing of these population parameters and, instead, rely on a statistical test in order to ascertain these differences.

Example of a one sample t-tests

Our objective here is to ascertain *whether there is a difference between the overall population's (company employees) break times and break times of employees in the engineering department.*

Let us now conduct a t-test at a 95% confidence level in order to find a difference (or not!). Technically speaking, this test will tell us if the sample comes from the same distribution as the population.

Assumptions of the one sample t-tests

Before diving into the five steps, we must first acknowledge that t-tests must satisfy the following two conditions to work properly:

- The population distribution should be normal, or the sample should be *large* ($n \geq 30$).
- In order to make the assumption that the sample is independently randomly sampled, it is sufficient to enforce that the population size should be at least 10 times larger than the sample size ($10n < N$).

Note that our test requires that either the underlying data be normal (which we know is not true for us), or that the sample size be at least 30 points large. For the t-test, this condition is sufficient to assume normality. This test also requires independence, which is satisfied by taking a sufficiently *small* sample. Sounds weird, right? The basic idea is that our sample must be large enough to assume normality (through conclusions similar to the central limit theorem) but small enough as to be *independent* from the population.

Now, let's follow our five steps:

1. Specify the hypotheses.

   We will let H_0 = *the engineering department takes breaks the same as the company as a whole*

   If we let this be the company average, we may write:

   $$H_0:$$

   Note how this is our *null*, or *default*, hypothesis. It is what we would assume, given no data. What we would like to show is the **alternative hypothesis**.

   Now that we actually have some options for our alternative, we could either say that the engineering mean (let's call it that) is lower than the company average, higher than the company average, or just flat out different (higher or lower) than the company average:

   - If we wish to answer the question, *is the sample mean different from the company average*, then this is called a **two-tailed test** and our alternative hypothesis would be as follows:

     $$Ha:$$

- If we want to answer *either is the sample mean lower than the company average or is the sample mean higher than the company average*, then we are dealing with a **one-tailed test** and our alternative hypothesis would be one or the other of the following hypotheses:

  Ha:(engineering takes longer breaks)

  Ha:(engineering takes shorter breaks)

The difference between one and two tails is the difference of dividing a number later on by 2 or not. The process remains completely unchanged for both. For this example, let's choose the two-tailed test. So, we are testing for whether or not this sample of the engineering department's average break times is different from the company average.

Our test will end in one of the two possible conclusions: we will either reject the null hypothesis, which means that the engineering department's break times are different from the company average, or we will fail to reject the null hypothesis, which means that there wasn't enough evidence in the sample to support rejecting the null.

2. Determine the sample size for the test sample.

   As mentioned earlier, most tests (including this one) make the assumption that either the underlying data is normal or that our sample is in the right range.

   - The sample is at least 30 points (it is 400)
   - The sample is less than 10% of the population (which would be 900 people)

3. Choose a significance level (usually called alpha or α).

   We will choose a 95% significance level, which means that our alpha would actually be *1 - .95 = .05*

4. Collect the data.

   Done! This was generated through the two Poisson distributions

5. Decide whether to reject or fail to reject the null hypothesis.

   As mentioned before, this step varies based on the test used. For a one sample t-test, we must calculate two numbers: the test statistic and our p value. Luckily, we can do this in one line in Python.

A test statistic is a value that is derived from sample data during a type of hypothesis test. They are used to determine whether or not to reject the null hypothesis.

The test statistic is used to compare the observed data with what is expected under the null hypothesis. The test statistic is used in conjunction with the p-value.

The p-value is the probability that the observed data occurred this way by chance.

When the data is showing very strong evidence against the null hypothesis, the test statistic becomes large (either positive or negative) and the p-value usually becomes very small, which means that our test is showing powerful results and what is happening is, probably, not happening by chance.

In the case of a t-test, a *t value* is our test statistic, as shown:

```
t_statistic, p_value = stats.ttest_1samp(a= engineering_breaks,
popmean= breaks.mean())
```

We input the `engineering_breaks` variable (which holds 400 break times) and the population mean, and we obtain the following numbers:

```
t_statistic == -5.742
p_value == .00000018
```

The test result shows that the *t value* is `-5.742`. This is a standardized metric that reveals the deviation of the sample mean from the null hypothesis. The p value is what gives us our final answer. Our p-value is telling us how often our result would appear by chance. So, for example, if our p-value was .06, then that would mean we should expect to observe this data by chance about 6% of the time. This means that about 6% of samples would yield results like this.

We are interested in how our p-value compares to our significance level:

- If the p-value is *less than* the significance level, then we can *reject* the null hypothesis
- If the p-value is *greater than* the significance level, then we *failed to reject* the null hypothesis

Our p value is way lower than .05 (our chosen significance level), which means that we may reject our null hypothesis in favor for the alternative. This means that the engineering department seems to take different break lengths than the company as a whole!

The use of p-values is controversial. Many journals have actually banned the use of p-values in tests for significance. This is because of the nature of the value. Suppose our p-value came out to .04. It means that 4% of the time, our data just randomly happened to appear this way and is not significant in any way. 4% is not that small of a percent! For this reason, many people are switching to different statistical tests. However, that does not mean that p-values are useless. It merely means that we must be careful and aware of what the number is telling us.

There are many other types of t-tests, including one-tailed tests (mentioned before) and paired tests as well as two sample t-tests (both not mentioned yet). These procedures can be readily found in statistics literature; however, we should look at something very important—what happens when we get it wrong.

Type I and type II errors

We've mentioned both the type I and type II errors in a previous chapter about probability in the examples of a binary classifier, but they also apply to hypothesis tests.

A type I error occurs if we reject the null hypothesis when it is actually true. This is also known as a **false positive**. The type I error rate is equal to the significance level α, which means that if we set a higher confidence level, for example, a significance level of 99%, our α is .01, and therefore our false positive rate is 1%.

A type II error occurs if we fail to reject the null hypothesis when it is actually false. This is also known as a **false negative**. The higher we set our confidence level, the more likely we are to actually see a type II error.

Hypothesis test for categorical variables

T-tests (among other tests) are hypothesis tests that work to compare and contrast quantitative variables and underlying population distributions. In this section, we will explore two new tests, both of which serve to explore qualitative data. They both are a form of test called chi-square tests. These two tests will perform the following two tasks for us:

- Determine whether a sample of categorical variables is taken from a specific population (similar to the t-test)
- Determine whether two variables affect each other and are associated to each other.

Chi-square goodness of fit test

The one-sample t-test was used to check whether a sample mean differed from the population mean. The chi-square goodness of fit test is very similar to the one sample t-test in that it tests whether the distribution of the sample data matches an expected distribution, while the big difference is that it is testing for categorical variables.

For example, a chi-square goodness of fit test would be used to see if the race demographics of your company match that of the entire city of the U.S. population. It can also be used to see if users of your website show similar characteristics to average Internet users.

As we are working with categorical data, we have to be careful because categories like "male", "female," or "other" don't have any mathematical meaning. Therefore, we must consider counts of the variables rather than the actual variables themselves.

In general, we use the chi-square goodness of fit test in the following cases:

- We want to analyze one categorical variable from one population
- We want to determine if a variable fits a specified or expected distribution

In a chi-square test, we compare what is observed to what we expect.

Assumptions of the chi-square goodness of fit test

There are two usual assumptions of this test, as follows:

- All the expected counts are at least 5
- Individual observations are independent and the population should be at least 10 times as large as the sample, ($10n < N$)

The second assumption should look familiar to the t-test; however, the first assumption should look foreign. Expected counts are something we haven't talked about yet but are about to!

When formulating our null and alternative hypotheses for this test, we consider a default distribution of categorical variables. For example, if we have a die and we are testing whether or not the outcomes are coming from a fair die, our hypothesis might look as follows:

H_0: *The specified distribution of the categorical variable is correct.*

$$p1 = 1/6, \quad p2 = 1/6, \quad p3 = 1/6, \quad p4 = 1/6, \quad p5 = 1/6, \quad p6 = 1/6$$

Our alternative hypothesis is quite simple, as shown:

H_a : The specified distribution of the categorical variable is not correct. At least one of the *pi* values is not correct.

In the t-test, we used our test statistic (the t value) to find our p-value. In a chi-square test, our test statistic is, well, chi-square.

Test Statistic: $\chi 2$ = *over k categories*

Degrees of Freedom = $k - 1$

A critical value is when we use $\chi 2$ as well as our degrees of freedom and our significance level, and then reject the null hypothesis if the p-value is below our significance level (the same as before).

Let's see an example to understand further.

Example of a chi-square test for goodness of fit

The CDC categorizes adult BMIs into four classes: Under/Normal, Over weight, Obesity, and Extreme Obesity. A 2009 survey showed the distribution for adults in the U.S. to be 31.2%, 33.1%, 29.4%, and 6.3% respectively. A total of 500 adults are randomly sampled and their BMI categories are recorded. Is there evidence to suggest that BMI trends have changed since 2009? Test at the 0.05 significance level.

	Under/Normal	Over	Obesity	Extreme Obesity	Total
Observed	102	178	186	34	500

First, let's calculate our expected values. In a sample of 500, we *expect* 156 to be Under/Normal (that's 31.2% of 500), and we fill in the remaining boxes in the same way.

	Under/Normal	Over	Obesity	Extreme Obesity	Total
Observed	102	178	186	34	500
Expected	156	165.5	147	31.5	500

First, check the conditions:

- All of the expected counts are greater than 5
- Each observation is independent and our population is very large (much more than 10 times of 500 people)

Next, carry out a goodness of fit test. We will set our null and alternative hypotheses:

- H_0: The 2009 BMI distribution is still correct.
- H_a: The 2009 BMI distribution is no longer correct (at least one of the proportions is different now). We can calculate our test statistic by hand:

$$\text{Test Statistic: } \chi^2 = \sum \frac{(Observed - Expected)^2}{Expected} \text{ for } df = 3$$

$$= \frac{(102-156)^2}{156} + \frac{(178-165.5)^2}{165.5} + \frac{(186-147)^2}{147} + \frac{(34-31.5)^2}{31.5} = 30.18$$

Alternatively, we can use our handy dandy Python skills, as shown:

```
observed = [102, 178, 186, 34]
expected = [156, 165.5, 147, 31.5]

chi_squared, p_value = stats.chisquare(f_obs= observed, f_exp=
expected)

chi_squared, p_value
#(30.1817679275599, 1.26374310311106e-06)
```

Our p-value is lower than .05; therefore, we may reject the null hypothesis in favor of the fact that the BMI trends today are different from what they were in 2009.

Chi-square test for association/independence

Independence as a concept in probability is when knowing the value of one variable tells you nothing about the value of another. For example, we might expect that the country and the month you were born in are independent. However, knowing which type of phone you use might indicate your creativity levels. Those variables might not be independent.

The chi-square test for association/independence helps us ascertain whether two categorical variables are independent of one another. The test for independence is commonly used to determine whether variables like education levels or tax brackets vary based on demographic factors, such as gender, race, and religion. Let's look back at an example posed in the preceding chapter, the A/B split test.

Recall that we ran a test and exposed half of our users to a certain landing page (Website A), exposed the other half to a different landing page (Website B), and then, measured the sign up rates for both sites. We obtained the following results:

	Did not sign up	**Signed up**
Website A	134	54
Website B	110	48

Results of our A/B test

We calculated website conversions but what we really want to know is whether there is a difference between the two variables: *which website was the user exposed to?* and *did the user sign up?*. For this, we will use our chi-square test.

Assumptions of the chi-square independence test

There are the following two assumptions of this test:

- All expected counts are at least 5
- Individual observations are independent and the population should be at least 10 times as large as the sample, ($10n < N$)

Note that they are exactly the same as the last chi-square test.

Let's set up our hypotheses:

- H_0 : There is no association between two categorical variables in the population of interest
- H_0 : Two categorical variables are independent in the population of interest
- H_a : There is an association between two categorical variables in the population of interest
- H_a : Two categorical variables are not independent in the population of interest

You might notice that we are missing something important here. Where are the expected counts? Earlier, we had a prior distribution to compare our observed results to but now we do not. For this reason, we will have to create some. We can use the following formula to calculate the expected values for each value. In each cell of the table, we can use:

Expected Count = to calculate our chi-square test statistic and our degrees of freedom

$$\text{Test Statistic: } \chi^2 = \sum \frac{(Observed_{r,c} - Expected_{r,c})^2}{Expected_{r,c}}$$
$$\text{over } r \text{ rows and } c \text{ columns}$$

$$\text{Degrees of Freedom} = (r-1) \cdot (c-1)$$

Here, r is the number of rows and c is the number of columns. Of course, as before, when we calculate our p-value, we will reject the null if that p-value is less than the significance level. Let's use some built-in Python methods, as shown, in order to quickly get our results:

```
observed = np.array([[134, 54],[110, 48]])
# built a 2x2 matrix as seen in the table above

chi_squared, p_value, degrees_of_freedom, matrix = stats.chi2_
contingency(observed= observed)

chi_squared, p_value
# (0.04762692369491045, 0.82724528704422262)
```

We can see that our p-value is quite large; therefore, we *fail* to reject our null hypothesis and we cannot say for sure that seeing a particular website has any effect on a user's sign up. There is no association between these variables.

Summary

In this chapter, we looked at different statistical tests, including chi-square and t-tests as well as point estimates and confidence intervals, in order to ascertain population parameters based on sample data. We were able to find that even with small samples of data, we can make powerful assumptions about the underlying population as a whole.

Statistics is a very wide and expansive subject that cannot truly be covered in a single chapter, however, our understanding of the subject will allow us to carry on and talk more about how we can use statistics and probability in order to communicate our ideas through data science in the next chapter.

9
Communicating Data

This chapter deals with the different ways of communicating results from our analysis. Here, we will look at different presentation styles as well as visualization techniques. The point of this chapter is to take our results and be able to explain them in a coherent, intelligible way so that anyone, whether they are data savvy or not, may understand and use our results.

Much of what we will discuss will be how to create effective graphs through labels, keys, colors, and more. We will also look at more advanced visualization techniques, such as parallel coordinate plots.

In this chapter, we will look into the following topics:

- Identifying effective and ineffective visualizations
- Recognizing when charts are attempting to "trick" the audience
- Being able to identify causation versus correlation
- Constructing appealing visuals that offer valuable insight

Why does communication matter?

Being able to conduct experiments and manipulate data in a coding language is not enough to conduct practical and applied data science. This is because data science is, generally, only as good as how it is used in practice. For instance, a medical data scientist might be able to predict the chance of a tourist contracting Malaria in developing countries with >98% accuracy, however, if these results are published in a poorly marketed journal and online mentions of the study are minimal, their groundbreaking results that could potentially prevent deaths would never see the true light of day.

For this reason, communication of results is arguably as important as the results themselves. A famous example of poor management of distribution of results is the case of Gregor Mendel. Mendel is widely recognized as one of the founders of modern genetics. However, his results (including data and charts) were not well adopted until after his death. Mendel even sent them to Charles Darwin, who largely ignored Mendel's papers, which were written in unknown Moravian journals.

Generally, there are two ways of presenting results: verbal and visual. Of course, both the verbal and visual forms of communication can be broken down into dozens of subcategories, including slide decks, charts, journal papers, and even university lectures. However, we can find common elements of data presentation that can make anyone in the field more aware and effective in their communication skills.

Let's dive right into effective (and ineffective) forms of communication, starting with visuals.

Identifying effective and ineffective visualizations

The main goal of data visualization is to have the reader quickly digest the data, including possible trends, relationships, and more. Ideally, a reader will not have to spend more than 5-6 seconds digesting a single visualization. For this reason, we must make visuals very seriously and ensure that we are making a visual as effective as possible. Let's look at four basic types of graphs: scatter plots, line graphs, bar charts, histograms, and box plots.

Scatter plots

A scatter plot is probably one of the simplest graphs to create. It is made by creating two *quantitative* axes and using data points to represent observations. The main goal of a scatter plot is to highlight relationships between two variables and, if possible, reveal a correlation.

For example, we can look at two variables: average hours of TV watched in a day and a 0-100 scale of work performance (0 being very poor performance and 100 being excellent performance). The goal here is to find a relationship (if it exists) between watching TV and average work performance.

The following code simulates a survey of a few people, in which they revealed the amount of television they watched, on an average, in a day against a company-standard work performance metric:

```
import pandas as pd
hours_tv_watched = [0, 0, 0, 1, 1.3, 1.4, 2, 2.1, 2.6, 3.2, 4.1, 4.4,
4.4, 5]
```

This line of code is creating 14 sample survey results of people answering the question of how many hours of TV they watch in a day.

```
work_performance = [87, 89, 92, 90, 82, 80, 77, 80, 76, 85, 80, 75,
73, 72]
```

This line of code is creating 14 new sample survey results of the same people being rated on their work performance on a scale from 0 to 100.

For example, the first person watched 0 hours of TV a day and was rated 87/100 on their work, while the last person watched, on an average, 5 hours of TV a day and was rated 72/100:

```
df = pd.DataFrame({'hours_tv_watched':hours_tv_watched, 'work_
performance':work_performance})
```

Here, we are creating a Dataframe in order to ease our exploratory data analysis and make it easier to make a scatter plot:

```
df.plot(x='hours_tv_watched', y='work_performance', kind='scatter')
```

Now, we are actually making our scatter plot. In the following plot, we can see that our axes represent the number of hours of TV watched in a day and the person's work performance metric:

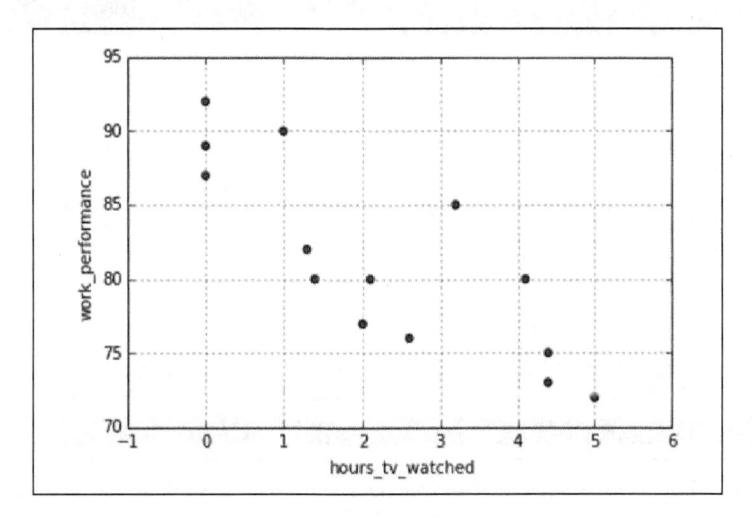

Each point on a scatter plot represents a single observation (in this case a person) and its location is a result of where the observation stands on each variable. This scatter plot does seem to show a relationship, which implies that as we watch more TV in the day, it seems to affect our work performance.

Of course, as we are now experts in statistics from the last two chapters, we know that this might not be causational. A scatter plot may only work to reveal a correlation or an association between but not a causation. Advanced statistical tests, such as the ones we saw in *Chapter 8, Advanced Statistics*, might work to reveal causation. Later on in this chapter, we will see the damaging effects that trusting correlation might have.

Line graphs

Line graphs are, perhaps, one of the most widely used graphs in data communication. A line graph simply uses lines to connect data points and usually represents time on the x axis. Line graphs are a popular way to show changes in variables over time. The line graph, like the scatter plot, is used to plot *quantitative* variables.

As a great example, many of us wonder about the possible links between what we see on TV and our behavior in the world. A friend of mine once took this thought to an extreme—he wondered if he could find a relationship between the TV show, The X-Files, and the amount of UFO sightings in the U.S.. He then found the number of sightings of UFOs per year and plotted them over time. He then added a quick graphic to ensure that readers would be able to identify the point in time when the X-files were released:

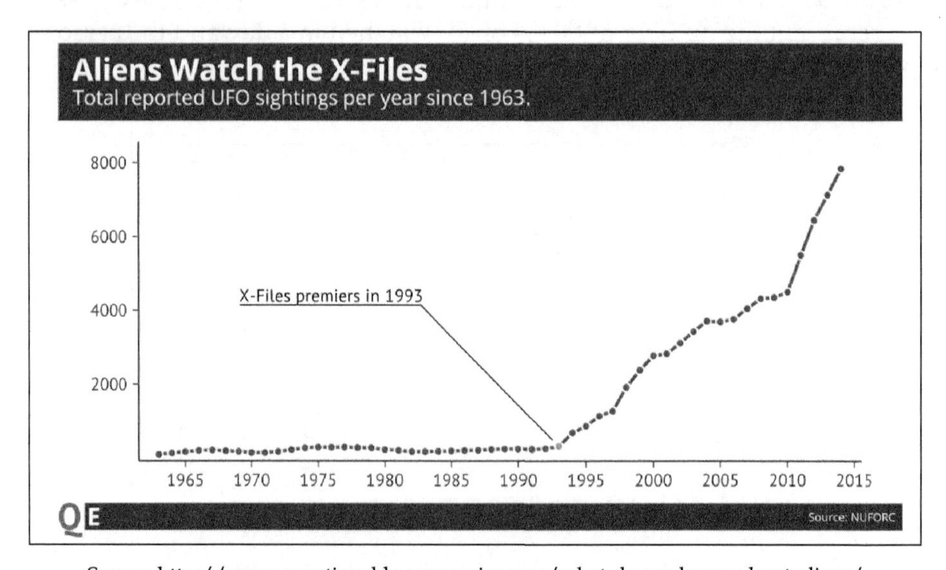

Source: http://www.questionable-economics.com/what-do-we-know-about-aliens/

It appears to be clear that right after 1993, the year of the X-Files premier, the number of UFO sightings started to climb drastically.

This graphic, albeit light-hearted, is an excellent example of a simple line graph. We are told what each axis measures, we can quickly see a general trend in the data, and we can identify with the author's intent, which is to show a relationship between the number of UFO sightings and the X-files premier.

On the other hand, the following is a less impressive line chart:

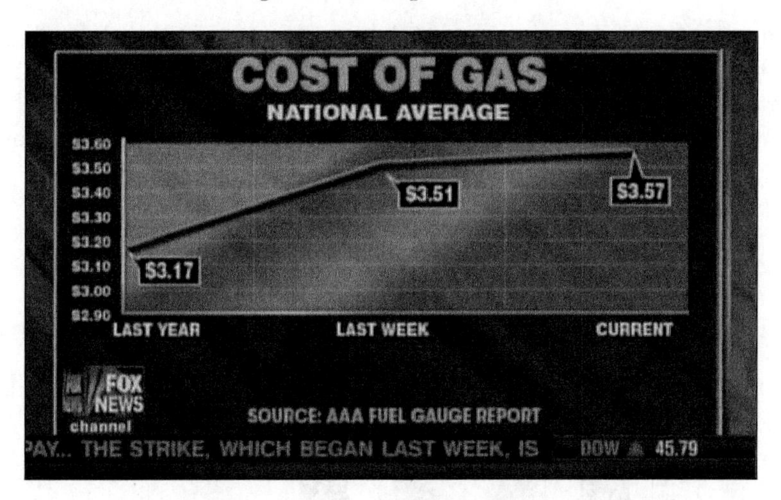

This line graph attempts to highlight the change in the price of gas by plotting three points in time. At first glance, it is not much different than the previous graph—we have time on the bottom *x* axis and a quantitative value on the vertical *y* axis. The (not so) subtle difference here is that the three points are equally spaced out on the *x* axis; however, if we read their actual time indications, they are not equally spaced out in time. A year separates the first two points whereas a mere 7 days separates the last two points.

Bar charts

We generally turn to bar charts when trying to compare variables across different groups. For example, we can plot the number of countries per continent using a bar chart. Note how the *x* axis does not represent a quantitative variable, in fact, when using a bar chart, the *x* axis is generally a categorical variable, while the *y* axis is quantitative.

Note that, for this code, I am using the World Health Organization's report on alcohol consumption around the world by country:

```
drinks = pd.read_csv('data/drinks.csv')

drinks.continent.value_counts().plot(kind='bar', title='Countries per
Continent')
plt.xlabel('Continent')
plt.ylabel('Count')
```

The following graph shows us a count of the number of countries in each continent. We can see the continent code at the bottom of the bars and the bar height represents the number of countries we have in each continent. For example, we see that Africa has the most countries represented in our survey, while South America has the least:

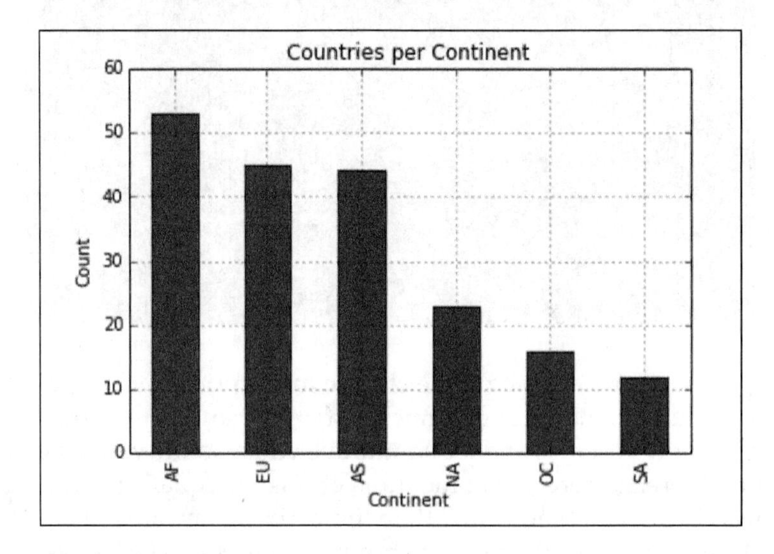

In addition to the count of countries, we can also plot the average beer servings per continent using a bar chart, as shown:

```
drinks.groupby('continent').beer_servings.mean().plot(kind='bar')
```

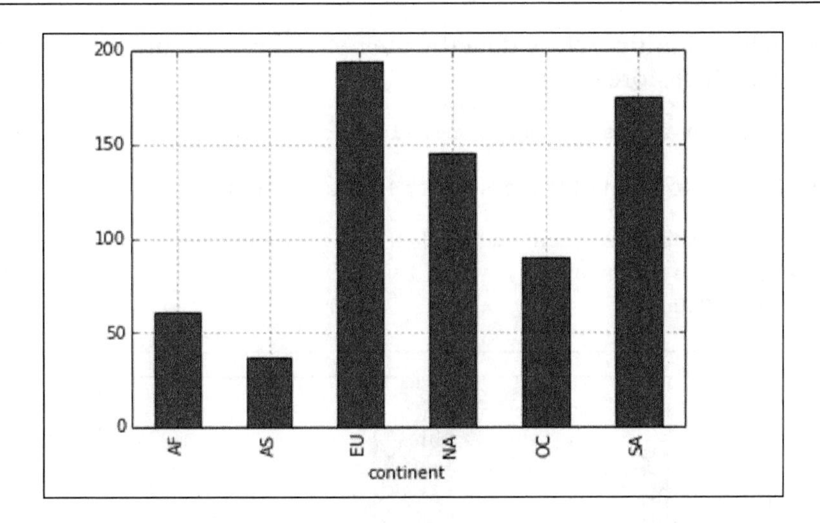

Note how a scatter plot or a line graph would not be able to support this data because they can only handle quantitative variables; bar graphs have the ability to demonstrate categorical values.

We can also use bar charts to graph variables that change over time, like a line graph.

Histograms

Histograms show the frequency distribution of a single quantitative variable by splitting up the data, by range, into equidistant *bins* and plotting the raw count of observations in each bin. A histogram is effectively a bar chart where the *x* axis is a bin (subrange) of values and the *y* axis is a count. As an example, I will import a store's daily number of unique customers, as shown:

```
rossmann_sales = pd.read_csv('data/rossmann.csv')

rossmann_sales.head()
```

	Store	DayOfWeek	Date	Sales	Customers	Open	Promo	StateHoliday	SchoolHoliday
0	1	5	2015-07-31	5263	555	1	1	0	1
1	2	5	2015-07-31	6064	625	1	1	0	1
2	3	5	2015-07-31	8314	821	1	1	0	1
3	4	5	2015-07-31	13995	1498	1	1	0	1
4	5	5	2015-07-31	4822	559	1	1	0	1

Note how we have multiple store data (by the first `Store` column). Let's subset this data for only the first store, as shown:

```
first_rossmann_sales = rossmann_sales[rossmann_sales['Store']==1]
```

Now, let's plot a histogram of the first store's customer count:

```
first_rossmann_sales['Customers'].hist(bins=20)
plt.xlabel('Customer Bins')
plt.ylabel('Count')
```

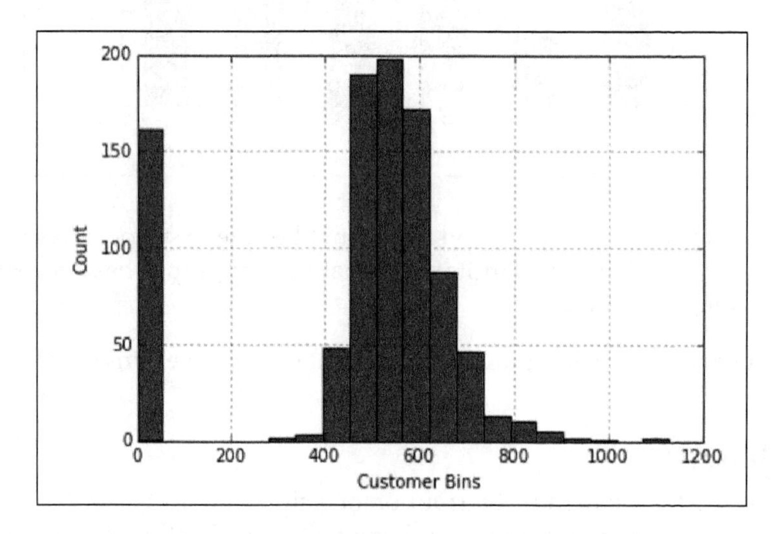

The *x* axis is now categorical in that each category is a selected range of values, for example, 600-620 customers would potentially be a category. The *y* axis, like a bar chart, is plotting the number of observations in each category. In this graph, for example, one might take away the fact that most of the time, the number of customers on any given day will fall between 500 and 700.

Altogether, histograms are used to visualize the distribution of values that a quantitative variable can take on.

In a histogram, we do not put spaces between bars.

Box plots

Box plots are also used to show a distribution of values. They are created by plotting the five number summary, as follows:

- The minimum value
- The first quartile (the number that separates the 25% lowest values from the rest)
- The median
- The third quartile (the number that separates the 25% highest values from the rest)
- The maximum value

In Pandas, when we create box plots, the red line denotes the median, the top of the box (or the right if it is horizontal) is the third quartile, and the bottom (left) part of the box is the first quartile.

The following is a series of box plots showing the distribution of beer consumption according to continents:

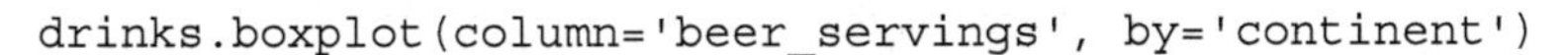

```
drinks.boxplot(column='beer_servings', by='continent')
```

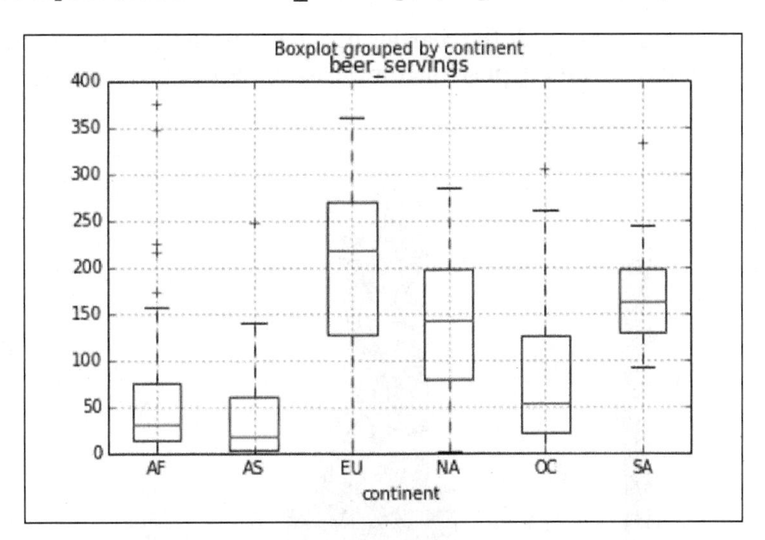

Now, we can clearly see the distribution of beer consumption across the seven continents and how they differ. Africa and Asia have a much lower median of beer consumption than Europe or North America.

Box plots also have the added bonus of being able to show outliers much better than a histogram. This is because the minimum and maximum are parts of the box plot.

Getting back to the customer data, let's look at the same store customer numbers, but using a box plot:

```
first_rossmann_sales.boxplot(column='Customers', vert=False)
```

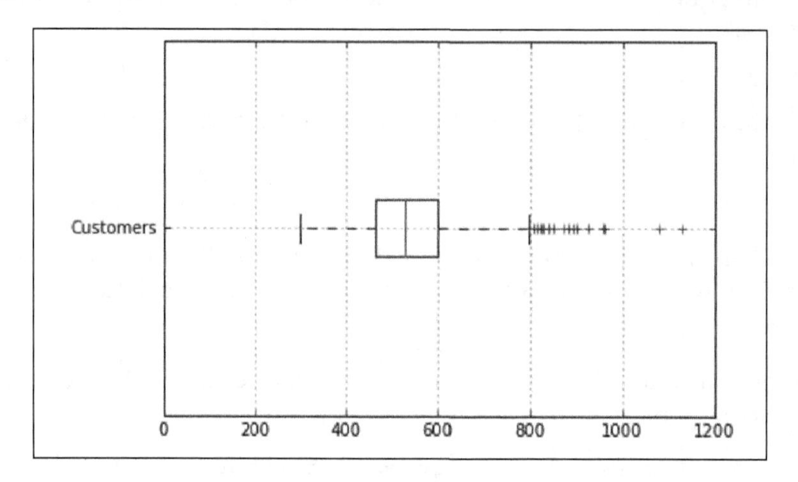

This is the exact same data as plotted earlier in the histogram; however, now it is shown as a box plot. For the purpose of comparison, I will show you both the graphs one after the other:

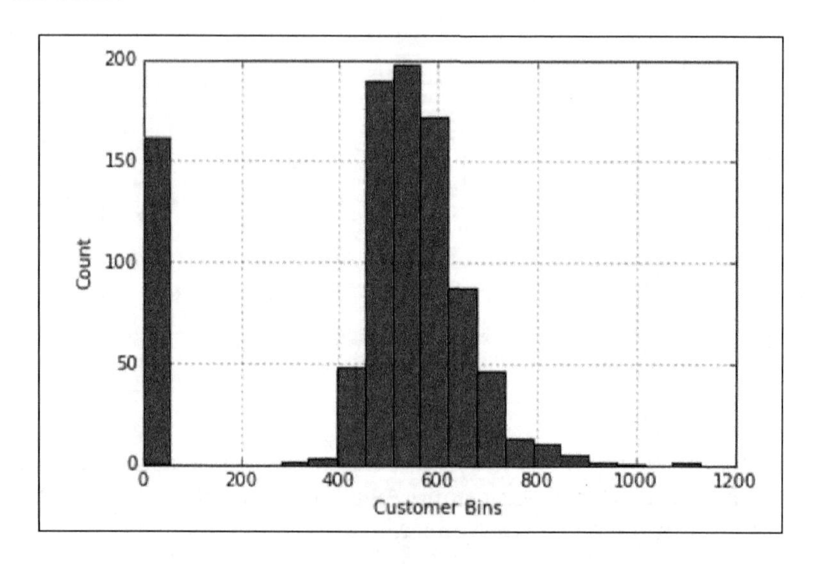

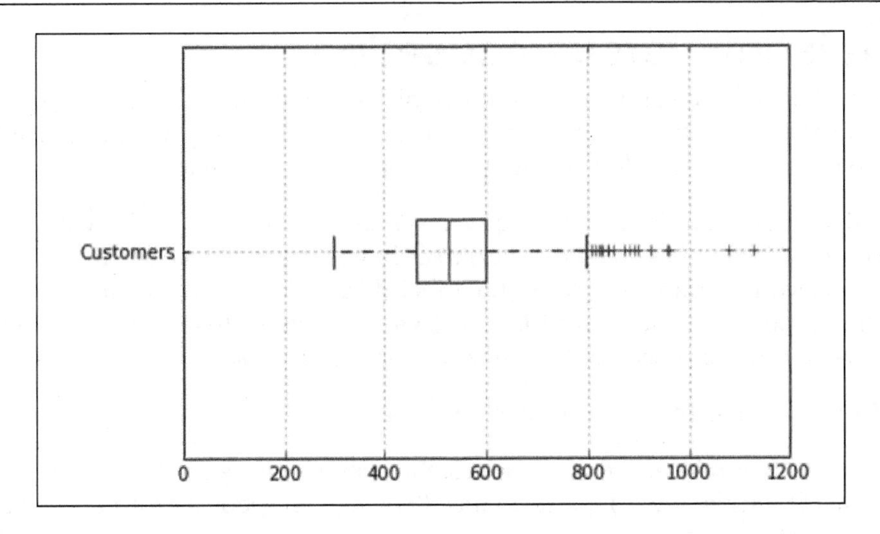

Note how the x axis for each graph are the same, ranging from 0 to 1,200. The box plot is much quicker at giving us a center of the data, the red line is the median, while the histogram works much better in showing us how spread out the data is and where people's biggest bins are. For example, the histogram reveals that there is a very large bin of zero people. This means that for a little over 150 days of data, there were zero customers.

Note that we can get the exact numbers to construct a box plot using the `describe` feature in Pandas, as shown:

```
first_rossmann_sales['Customers'].describe()
```

```
min         0.000000
25%       463.000000
50%       529.000000
75%       598.750000
max      1130.000000
```

When graphs and statistics lie

I should be clear, statistics don't lie, people lie. One of the easiest ways to trick your audience is to confuse correlation with causation.

Correlation versus causation

I don't think I would be allowed to publish this book without taking a deeper dive into the differences between correlation and causation. For this example, I will continue to use my data of TV consumption and work performance.

Correlation is a quantitative metric between -1 and 1 that measures how two variables *move with each other*. If two variables have a correlation close to -1, it means that as one variable increases, the other decreases, and if two variables have a correlation close to +1, it means that those variables move together in the same direction—as one increases, so does the other, and vice versa.

Causation is the idea that one variable affects another.

For example, we can look at two variables: the average hours of TV watched in a day and a 0-100 scale of work performance (0 being very poor performance and 100 being excellent performance). One might expect that these two factors are negatively correlated, which means that as the number of hours of TV watched increases in a 24 hour day, your overall work performance goes down. Recall the code from earlier, which is as follows:

```
import pandas as pd
hours_tv_watched = [0, 0, 0, 1, 1.3, 1.4, 2, 2.1, 2.6, 3.2, 4.1, 4.4,
4.4, 5]
```

Here, I am looking at the same sample of 14 people as before and their answers to the question, *how many hours of TV do you watch on average per night*:

```
work_performance = [87, 89, 92, 90, 82, 80, 77, 80, 76, 85, 80, 75,
73, 72]
```

These are the same 14 people as mentioned earlier, in the same order, but now, instead of the number of hours they watched TV, we have their work performance as graded by the company or a third-party system:

```
df = pd.DataFrame({'hours_tv_watched':hours_tv_watched, 'work_
performance':work_performance})
```

Earlier, we looked at a scatter plot of these two variables and it seemed to clearly show a downward trend between the variables—as TV consumption went up, work performance seemed to go down. However, a correlation coefficient, a number between -1 and 1, is a great way to identify relationships between variables and, at the same time, quantify them and categorize their strength.

Now we can introduce a new line of code that shows us the correlation between these two variables:

```
df.corr() # -0.824
```

Recall that a correlation, if close to -1, implies a strong negative correlation, while a correlation close to +1 implies a strong positive correlation.

This number helps support the hypothesis because a correlation coefficient close to -1 implies not only a negative correlation, but a strong one at that. Again, we can see this via a scatter plot between the two variables. So, both our visual and our numbers are aligned with each other. This is an important concept that should be true when communicating results. If your visuals and your numbers are off, people are less likely to take your analysis seriously:

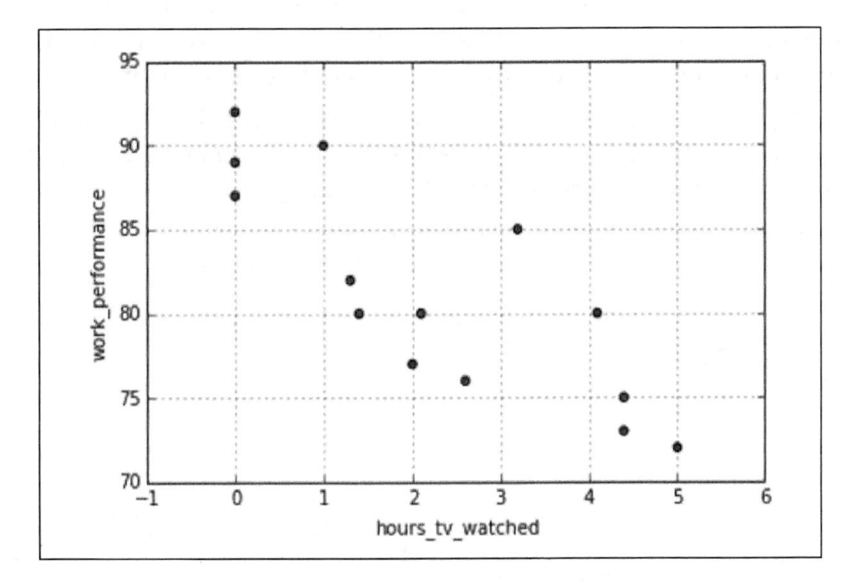

I cannot stress enough that correlation and causation are *different* from each other. Correlation simply quantifies the degree to which variables change together, whereas causation is the idea that one variable actually determines the value of another. If you wish to share the results of your findings of your correlational work, you might be met with challengers in the audience asking for more work to be done. What is more terrifying is that no one might know that the analysis is incomplete and you may make actionable decisions based on simple correlational work.

It is very often the case that two variables might be correlated to each other but they do not have any causation between them. This can be for a variety of reasons, some of which are as follows:

- There might be a *confounding factor* between them. This means that there is a third lurking variable that is not being factored and that acts as a bridge between the two variables. For example, previously, we showed that you might find that the amount of TV you watch is negatively correlated with work performance, that is, as the number of hours of TV you watch increases, your overall work performance may decrease. That is a correlation. It doesn't seem quite right to suggest that watching TV is the actual cause of a decrease in the quality of work performance. It might seem more plausible to suggest that there is a third factor, perhaps hours of sleep every night, that might answer this question. Perhaps, watching more TV decreases the amount of time you have for sleep, which in turn limits your work performance. The number of hours of sleep per night is the confounding factor.
- They might not have anything to do with each other! It might simply be a coincidence. There are many variables that are correlated but simply do not cause each other. Consider the following example:

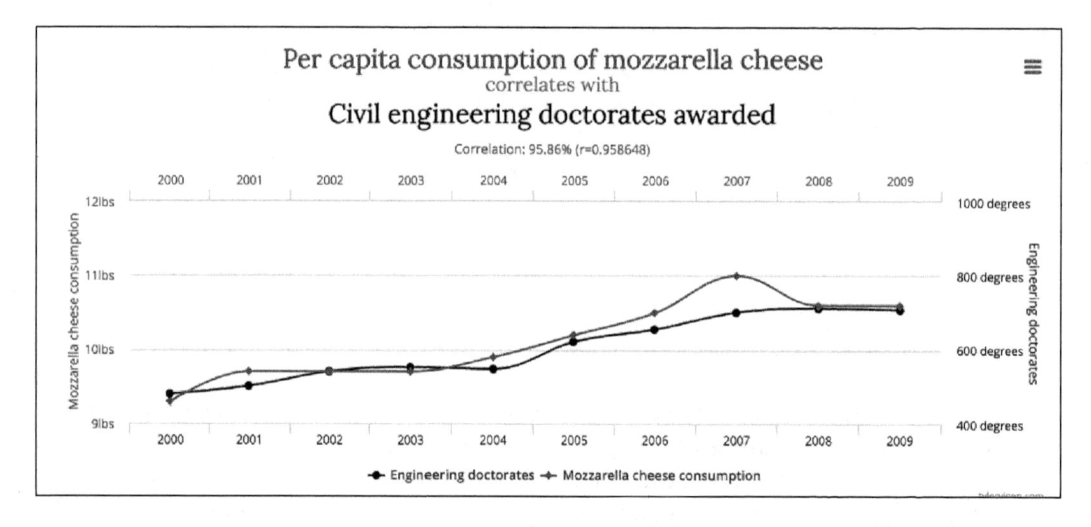

It is much more likely that these two variables only happen to correlate (more strongly than our previous example, I may add) that cheese consumption determines the number of civil engineering doctorates in the world.

You have likely heard the statement *correlation does not imply causation* and the last graph is exactly the reason why data scientists must believe that. Just because there exists a mathematical correlation between variables does not mean they have causation between them. There might be confounding factors between them or they just might not have anything to do with each other!

Let's see what happens when we ignore confounding variables and correlations become extremely misleading.

Simpson's paradox

Simpson's paradox is a formal reason for why we need to take confounding variables seriously. The paradox states that a correlation between two variables can be completely reversed when we take different factors into account. This means that even if a graph might show a positive correlation, these variables can become *anti-correlated* when another factor (most likely a confounding one) is taken into consideration. This can be very troublesome to statisticians.

Suppose we wish to explore the relationship between two different splash pages (recall our previous A/B testing in *Chapter 7, Basic Statistics*). We will call these pages page A and page B once again. We have two splash pages that we wish to compare and contrast and our main metric for choosing will be in our conversion rates, just as earlier.

Suppose we run a preliminary test and find the following conversion results:

Page A	**Page B**
75% (263/350)	83% (248/300)

This means that page B has almost a 10% higher conversion rate than page A. So, right off the bat, it seems like page B is the better choice because it has a higher rate of conversion. If we were going to communicate this data to our colleagues, it would seem that we are in the clear!

However, let's see what happens when we also take into account the coast that the user was closer to, as shown:

	Page A	**Page B**
West Coast	**95% (76 / 80)**	93% (231/250)
East Coast	**72% (193/270)**	34% (17/50)
Both	75% (263/350)	**83% (248/300)**

Thus the paradox! When we break the sample down by location, it seems that Page A was better in *both* categories but was worse overall. That's the beauty and, also, the horrifying nature of the paradox. This happens because of the unbalanced classes between the four groups.

The Page A / East Coast group as well as the Page B / West Coast group are providing most of the people in the sample, therefore skewing the results to be non-expected. The confounding variable here might be the fact that the pages were given at different hours of the day and the west coast people were more likely to see page B, while the East coast people were more likely to see page A.

There is a resolution to Simpson's paradox (and therefore an answer), however, the proof lies in a complex system of Bayesian networks and is a bit out of the scope of this book.

The main takeaway from Simpson's paradox is that we should not unduly give causational power to correlated variables. There might be confounding variables that have to be examined. Therefore, if you are able to reveal a correlation between two variables (such as website category and conversation rate or TV consumption and work performance), then you should absolutely try to isolate as many variables as possible that might be the reason for the correlation or can at least help explain your case further.

If correlation doesn't imply causation, then what does?

As a data scientist, it is often quite frustrating to work with correlations and not be able to draw conclusive causality. The best way to confidently obtain causality is, usually, through randomized experiments, such as the ones we saw in *Chapter 8, Advanced Statistics*. One would have to split up the population groups into randomly sampled groups and run hypothesis tests to conclude, with a degree of certainty, that there is a true causation between variables.

Verbal communication

Apart from visual demonstrations of data, verbal communication is just as important when presenting results. If you are not merely uploading results or publishing, you are usually presenting data to a room of data scientists, executives, or to a conference hall.

In any case, there are key areas to focus on when giving a verbal presentation, especially when the presentation is regarding findings about data.

There are generally two styles of oral presentations: one meant for more professional settings, including corporate offices where the problem at hand is usually tied directly to company performance or some other **KPI** (**key performance indicator**), and another meant more for a room of your peers where the key idea is to motivate the audience to care about your work.

It's about telling a story

Whether it is a formal or casual presentation, people like to hear stories. When you are presenting results, you are not just spitting out facts and metrics, you are attempting to frame the minds of your audience to believe in and care about what you have to say.

When giving a presentation, always be aware of your audience and try to gauge their reactions/interest in what you are saying. If they seem unengaged, try to relate the problem to them:

"Just think, when popular TV shows like Game of Thrones come back, your employees will all spend more time watching TV and therefore will have a lower work performance."

Now you have their attention. It's about relating to your audience, whether it's your boss or your mom's friend; you have to find a way to make it relevant.

On the more formal side of things

When presenting data findings to a more formal audience, I like to stick to the following six steps:

1. Outline the state of the problem.

   In this step, we go over the current state of the problem, including what the problem is and how the problem came to the attention of the team of data scientists.

2. Define the nature of the data.

   Here, we go more in depth about who this problem affects, how the solution would change the situation, and previous work done on the problem, if any.

3. Divulge an initial hypothesis.

   Here, we state what we believed to be the solution before doing any work. This might seem like a more novice approach to presentations; however, this can be a good time to outline not just your initial hypothesis, but, perhaps, the hypothesis of the entire company. For example, "we took a poll and 61% of the company believes there is no correlation between hours of TV watched and work performance".

4. Describe the solution and, possibly, the tools that led to the solution.

   Get into how you solved the problem, any statistical tests used, and any assumptions that were made during the course of the problem.

5. Share the impact that your solution will have on the problem.

   Talk about whether your solution was different from the initial hypothesis. What will this mean for the future? How can we take action on this solution to improve ourselves and our company?

6. Future steps.

   Share what future steps can be taken with the problem, such as how to implement the said solution and what further work this research sparked.

By following these steps, we can hit on all of the major areas of the data scientific method. The first thing you want to hit on during a formal presentation is action. You want your words and solutions to be actionable. There must be a clear path to take upon the completion of the project and the future steps should be defined.

The why/how/what strategy of presenting

When speaking on a less formal level, the why/how/what strategy is a quick and easy way to create a presentation worthy of praise. It is quite simple, as shown:

1. Tell your audience why this question is important without really getting into what you are actually doing.
2. Then, get into how you tackled this problem, using data mining, data cleaning, hypothesis testing, and so on.
3. Finally, tell them what your outcomes mean for the audience.

This model is borrowed from famous advertisements. The kind where they would not even tell you what the product was until 3 seconds left. They want to catch your attention and then, finally, reveal what it was that was so exciting. Consider the following example:

"Hello everyone, I am here to tell you about why we seem to have a hard time focusing on our job when the Olympics are being aired. After mining survey results and merging this data with company-standard work performance data, I was able to find a correlation between the number of hours of TV watched per day and average work performance. Knowing this, we can all be a bit more aware of our TV watching habits and make sure we don't let it affect our work. Thank you."

This chapter was actually formatted in this way! We started with *why* we should care about data communication, then we talked about *how* to accomplish it (through correlation, visuals, and so on), and finally, I am telling you the what, *which* is the why/how/what strategy (insert mind blowing sound effect here).

Summary

Data communication is not an easy task. It is one thing to understand the mathematics of how data science works, but it is a completely different thing to try to convince a room of data scientists and non-data scientists alike of your results and their value to them. In this chapter, we went over basic chart making as well as how to identify faulty causation and the ability to hone our oral presentation skills.

Our next few chapters will really begin to hit at one of the biggest talking points of data science. In the last nine chapters, we spoke about everything between how to obtain data, clean data, and visualize data in order to gain a better understanding of the environment that the data represents.

We then turned to looking at the basic and advanced probability/statistics laws in order to use quantifiable theorems and tests on our data to get actionable results and answers.

In subsequent chapters, we will take a look into machine learning and the nature in which machine learning performs well and doesn't perform well. As we take a journey into this material, I urge you, the reader, to keep an open mind and truly understand not just how machine learning works, but also why we need to use it.

10

How to Tell If Your Toaster Is Learning – Machine Learning Essentials

Machine learning has become quite the phrase of the decade. It seems as though every time we hear about the next greatest startup or turn on the news, we hear something about a revolutionary piece of machine learning technology and how it will change the way we live.

This chapter focuses on machine learning as a practical part of data science. We will cover the following topics in this chapter:

- Defining the different types of machine learning, along with examples of each kind
- Areas in regression, classification, and more
- What is machine learning and how it is used in data science
- The differences between machine learning and statistical modeling and how machine learning is a broader category of the latter

Our aim will be to utilize statistics, probability, and algorithmic thinking in order to understand and apply essential machine learning skills to practical industries, such as marketing. Examples will include predicting star ratings of restaurant reviews, predicting the presence of a disease, spam e-mail detection, and much more. This chapter focuses more on machine learning as a whole and a single statistical model. The subsequent chapters will deal with many more models, some of which are much more complex.

We will also turn our focus on metrics, which tell us how effective our models are. We will use metrics in order to conclude results and make predictions using machine learning.

What is machine learning?

It wouldn't make sense to continue without a concrete definition of what machine learning is. Well, let's back up for a minute. In *Chapter 1, How to Sound Like a Data Scientist,* we defined machine learning as giving computers the ability to learn from data without being given explicit *rules* by a programmer. This definition still holds true. Machine learning is concerned with the ability to ascertain certain patterns (signals) out of data, even if the data has inherent errors in it (noise).

Machine learning models are able to learn from data without the explicit help of a human. That is the main difference between machine learning models and *classical algorithms*. Classical algorithms are told how to find the best answer in a complex system and the algorithm then searches for these best solutions and often works faster and more efficiently than a human. However, the bottleneck here is that the human has to first come up with the *best solution*. In machine learning, the model is not told the best solution and instead, is given several examples of the problem and is told, *figure out the best solution*.

Machine learning is just another tool in the tool belt of a data scientist. It is on the same level as statistical tests (chi-square or t-tests) or uses base probability/statistics to estimate population parameters. Machine learning is often regarded as the only thing data scientists know how to do, and this is simply untrue. A true data scientist is able to recognize when machine learning is applicable and more importantly, when it is not.

Machine learning is a game of correlations and relationships. Most machine learning algorithms in existence are concerned with finding and/or exploiting relationships between datasets (often represented as columns in a Pandas Dataframe). Once machine learning algorithms can pinpoint on certain correlations, the model can either use these relationships to predict future observations or generalize the data to reveal interesting patterns.

Perhaps a great way to explain machine learning is to offer an example of a problem coupled with two possible solutions: one using a machine learning algorithm and the other utilizing a non-machine learning algorithm.

Example – facial recognition

This problem is very well documented. Given a picture of a face, whose face does it belong to? However, I argue that there is a more important question that must be asked even before this. Suppose you wish to implement a home security system that recognizes who is entering your house. Most likely, during the day, your house will be empty most of the time and the facial recognition must kick in only if there is a person in the shot. This is exactly the question I propose we try and solve—given a photo, is there a face in it?

Given this problem, I propose the following two solutions:

- The non-machine learning algorithm that will *define* a face as having a roundish structure, two eyes, hair, nose, and so on. The algorithm then looks for these *hard-coded* features in the photo and returns whether or not it was able to find any of these features.
- The machine learning algorithm will work a bit differently. The model will only be given several pictures of faces and non-faces that are labeled as such. From the examples (called training sets) it would figure out its own definition of a face.

The machine learning version of the solution is never told what a face is, it is merely given several examples, some with faces, and some without. It is then up to the machine learning model to figure out the difference between the two. Once it figures out the difference between the two, it uses this information to take in a picture and *predict* whether or not there is a face in the new picture. For example, to train the model, we will give it the following three images:

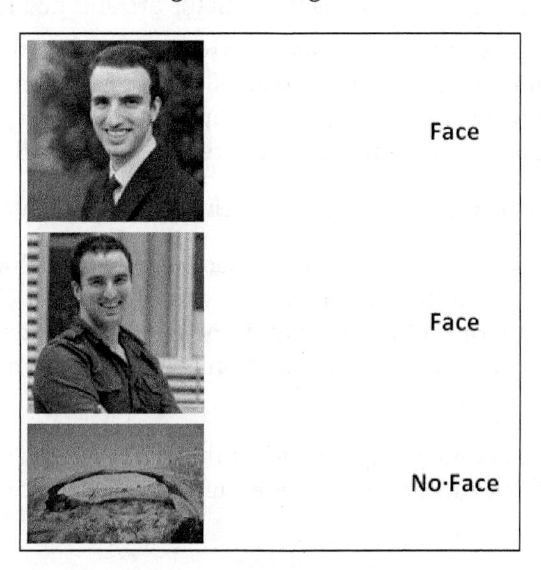

The model will then figure out the difference between the pictures labeled as **Face** and the images labeled as **No Face** and be able to use that difference to find faces in the future photos.

Machine learning isn't perfect

There are many caveats of machine learning. Many are specific to different models being implemented, but there are some assumptions that are universal for any machine learning model, as follows:

- The data used is, for the most part, is preprocessed and cleaned using the methods outlined in the earlier chapters.

  Almost no machine learning model will tolerate dirty data with missing values or categorical values. Use dummy variables and filling/dropping techniques to handle these discrepancies.

- Each row of a cleaned dataset represents a single observation of the environment we are trying to model.
- If our goal is to find relationships between variables, then there is an assumption that there is some kind of relationship between these variables.

  This assumption is particularly important. Many machine learning models take this assumption very seriously. These models are not able to communicate that there might not be a relationship.

- Machine learning models are generally considered semiautomatic, which means that intelligent decisions by humans are still needed.

  The machine is very smart but has a hard time putting things into context. The output of most models is a series of numbers and metrics attempting to quantify how well the model did. It is up to a human to put these metrics into perspective and communicate the results to an audience

- Most machine learning models are sensitive to noisy data.

  This means that the models get confused when you include data that doesn't make sense. For example, if you are attempting to find relationships between economic data around the world and one of your columns is puppy adoption rates in the capital city, that information is likely not to be relevant and will confuse the model.

These assumptions will come up again and again when dealing with machine learning. They are all too important and often ignored by novice data scientists.

How does machine learning work?

Each flavor of machine learning and each individual model works in very different ways, exploiting different parts of mathematics and data science. However, in general, machine learning works by taking in data, finding relationships within the data, and giving as output what the model learned, as illustrated in the following diagram:

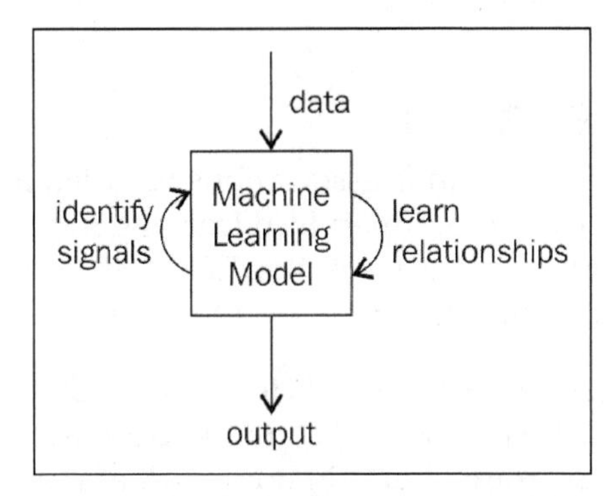

An overview of machine learning models

As we explore the different types of machine learning models, we will see how they manipulate data differently and come up with different outputs for different applications.

Types of machine learning

There are many ways to segment machine learning and dive deeper. In *Chapter 1, How to Sound Like a Data Scientist,* I mentioned statistical and probabilistic models. These models utilize statistics and probability, which we've seen in the previous chapters, in order to find relationships between data and make predictions. In this chapter, we will implement both types of models. In the following chapter, we will see machine learning outside the rigid mathematical world of statistics/probability. One can segment machine learning models by different characteristics, including:

- The types of data/organic structures they utilize (tree/graph/neural network)
- The field of mathematics they are most related to (statistical/probabilistic)
- The level of computation required to train (deep learning)

For the purpose of education, I will offer my own breakdown of machine learning models. Branching off of the top level of machine learning, there are the following three subsets:

- Supervised learning
- Unsupervised learning
- Reinforcement learning

Supervised learning

Simply put, supervised learning finds associations between features of a dataset and a target variable. For example, supervised learning models might try to find the association between a person's health features (heart rate, obesity level, and so on) and that person's risk of having a heart attack (the target variable).

These associations allow supervised models to make predictions based on past examples. This is often the first thing that comes to people's minds when they hear the phrase, machine learning, but it in no way does it encompass the realm of machine learning. Supervised machine learning models are often called **predictive analytics models**, named for their ability to predict the future based on the past.

Supervised machine learning requires a certain type of data called **labeled data**. This means that we must teach our model by giving it historical examples that are labeled with the correct answer. Recall the facial recognition example. That is a supervised learning model because we are training our model with the previous pictures labeled as either *face* or *not face*, and then asking the model to *predict* whether or not a new picture has a face in it.

Specifically, supervised learning works using parts of the data to predict another part. First, we must separate data into two parts, as follows:

- The predictors, which are the columns that will be used to make our prediction.

  These are sometimes called features, inputs, variables, and independent variables.

- The response, which is the column that we wish to predict.

  This is sometimes called outcome, label, target, and dependent variable.

Supervised learning attempts to find a relationship between the predictors and the response in order to make a prediction. The idea is that in the future a data observation will present itself and we will only know the predictors. The model will then have to use the predictors to make an accurate prediction of the response value.

Example - heart attack prediction

Suppose we wish to predict if someone will have a heart attack within a year. To predict this, we are given that person's cholesterol, blood pressure, height, their smoking habits, and perhaps more. From this data, we must ascertain the likelihood of a heart attack. Suppose, to make this prediction, we look at the previous patients and their medical history. As these are previous patients, we know not only their predictors (cholesterol, blood pressure, and so on), but we also know if they actually had a heart attack (because it already happened!).

This is a supervised machine learning problem because we are:

- Making a prediction about someone
- Using historical training data to find relationships between medical variables and heart attacks

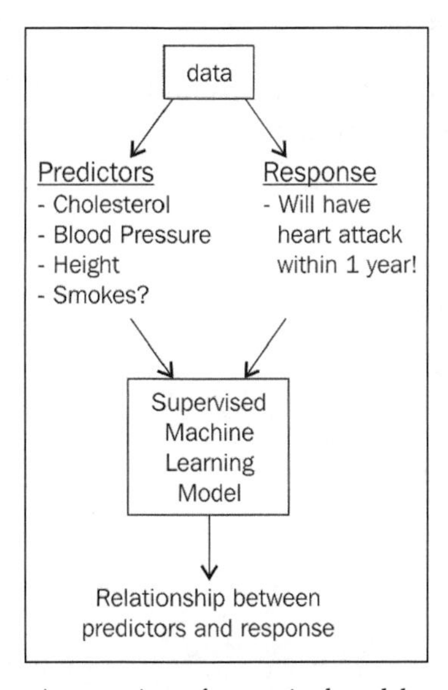

An overview of supervised models

The hope here is that a patient will walk in tomorrow and our model will be able to identify whether or not the patient is at risk for a heart attack based on her/his conditions (just like a doctor would!).

As the model sees more and more labeled data, it adjusts itself in order to match the correct labels given to us. We can use different metrics (explained later in this chapter) to pinpoint exactly how well our supervised machine learning model is doing and how it can better adjust itself.

One of the biggest drawbacks of supervised machine learning is that we need this labeled data, which can be very difficult to get a hold of. Suppose we wish to predict heart attacks, we might need thousands of patients along with all of their filled in medical information and years' worth of follow-up records for each person, which could be a nightmare to obtain.

In short, supervised models use historical labeled data in order to make predictions about the future. Some possible applications for supervised learning include:

- Stock price predictions
- Weather predictions
- Crime predictions

Note how each of the preceding examples uses the word **prediction**, which makes sense seeing how I emphasized supervised learning's ability to make predictions about the future. Predictions, however, are not where the story ends.

Here is a visualization of how supervised models use labeled data to *fit* themselves and prepare themselves to make predictions:

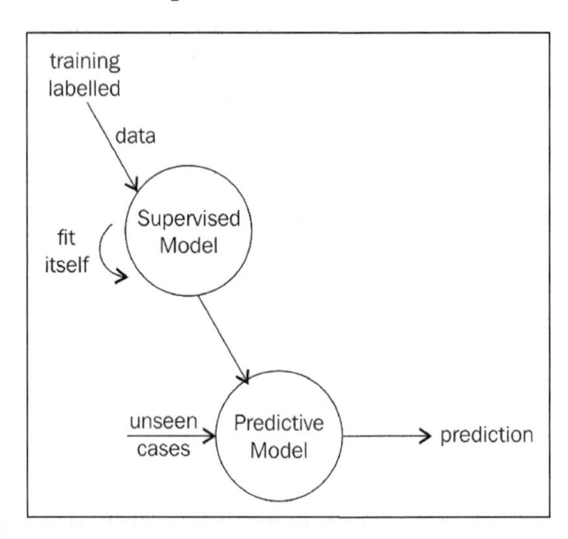

Note how the supervised model learns from a bunch of training data and then, when it is ready, it looks at unseen cases and outputs a prediction.

It's not only about predictions

Supervised learning exploits the relationship between the predictors and the response to make predictions, but sometimes, it is enough just knowing that there even is a relationship. Suppose we are using a supervised machine learning model to predict whether or not a customer will purchase a given item. A possible dataset might look as follows:

Person ID	Age	Gender	Employed?	Bought the product?
1	63	F	N	Y
2	24	M	Y	N

Note that, in this case, our predictors are Age, Gender, and Employed, while our response is Bought the product? This is because we want to see if, given someone's age, gender, and employment status, they will buy the product.

Assume that a model is trained on this data and can make accurate predictions about whether or not someone will buy something. That, in and of itself, is exciting but there's something else that is arguably even more exciting. The fact that we could make accurate predictions implies that there is a relationship between these variables, which means that to know if someone will buy your product, you only need to know their age, gender, and employment status! This might contradict the previous market research indicating that much more must be known about a potential customer to make such a prediction.

This speaks to supervised machine learning's ability *to understand which predictors affect the response and how*. For example, are women more likely to buy the product, which age groups are prone to decline the product, is there a combination of age and gender that is a better predictor than any one column on its own? As someone's age increases, do their chances of buying the product go up, down, or stay the same?

It is also possible that all the columns are not necessary. A possible output of a machine learning might suggest that only certain columns are necessary to make the prediction and that the other columns are only noise (they do not correlate to the response and therefore confuse the model).

Types of supervised learning

There are, in general, two types of supervised learning models: **regression** and **classification**. The difference between the two is quite simple and lies in the response variable.

Regression

Regression models attempt to predict a continuous response. This means that the response can take on a range of infinite values. Consider the following examples:

- Dollar amounts
    - Salary
    - Budget
- Temperature
- Time
    - Generally recorded in seconds or minutes

Classification

Classification attempts to predict a categorical response, which means that the response only has a finite amount of choices. Examples include the ones given as follows:

- Cancer grade (1, 2, 3, 4, 5)
- True/False questions, such as the following examples:
    - "Will this person have a heart attack within a year?"
    - "Will you get this job?"
- Given a photo of a face, who does this face belong to? (facial recognition)
- Predict the year someone was born:
    - Note that there are many possible answers (over 100) but still finitely many more

Example – regression

The following graphs show a relationship between three categorical variables (age, year they were born, and education level) and a person's wage:

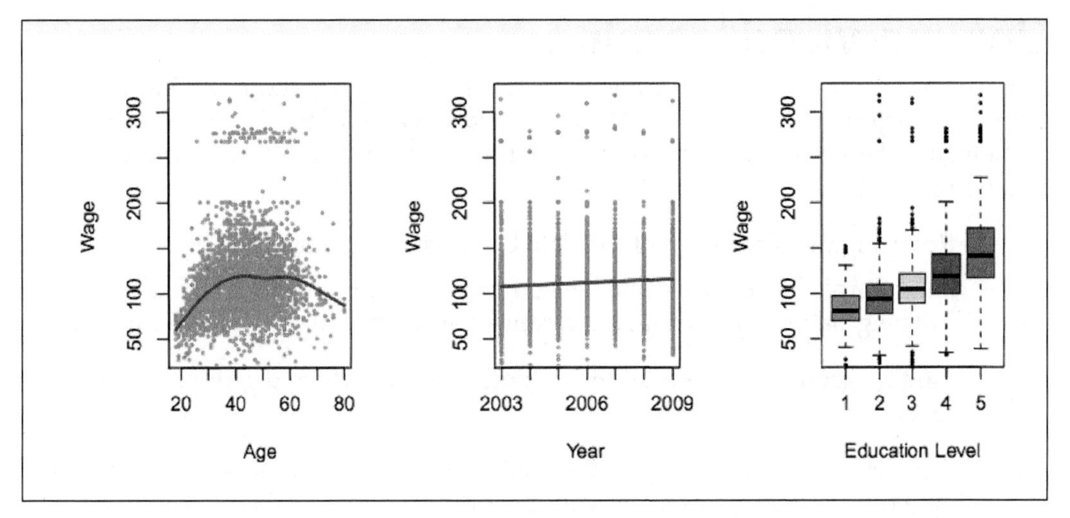

Source: https://lagunita.stanford.edu/c4x/HumanitiesScience/StatLearning/asset/introduction.pdf

Note that even though each predictor is categorical, this example is regressive because the *y* axis, our dependent variable, our response, is continuous.

Our earlier heart attack example is classification because the response was *will this person have a heart attack within a year?*, which has only two possible answers: *Yes* or *No*.

Data is in the eyes of the beholder

Sometimes, it can be tricky to decide whether or not you should use classification or regression. Consider that we are interested in the weather outside. We could ask the question, *how hot is it outside?*, in which case your answer is on a continuous scale, and some possible answers are 60.7 degrees, or 98 degrees. However, as an exercise, go and ask 10 people what the temperature is outside. I guarantee you that someone (if not most people) will not answer in some exact degrees but will bucket their answer and say something *like it's in the 60s.*

We might wish to consider this problem as a classification problem, where the response variable is no longer in exact degrees but is in a bucket. There would only be a finite number of buckets in theory, making the model perhaps learn the differences between 60s and 70s a bit better.

Unsupervised learning

The second type of machine learning does not deal with predictions but has a much more open objective. Unsupervised learning takes in a set of predictors and utilizes relationships between the predictors in order to accomplish tasks, such as the following:

- Reducing the dimension of the data by condensing variables together.

  An example of this would be file compression. Compression works by utilizing patterns in the data and representing the data in a smaller format.

- Finding groups of observations that behave similarly and grouping them together.

The first element on this list is called **dimension reduction** and the second is called **clustering**. Both of these are examples of unsupervised learning because they do not attempt to find a relationship between predictors and a specific response and therefore are not used to make predictions of any kind. Unsupervised models, instead, are utilized to find organizations and representations of the data that were previously unknown.

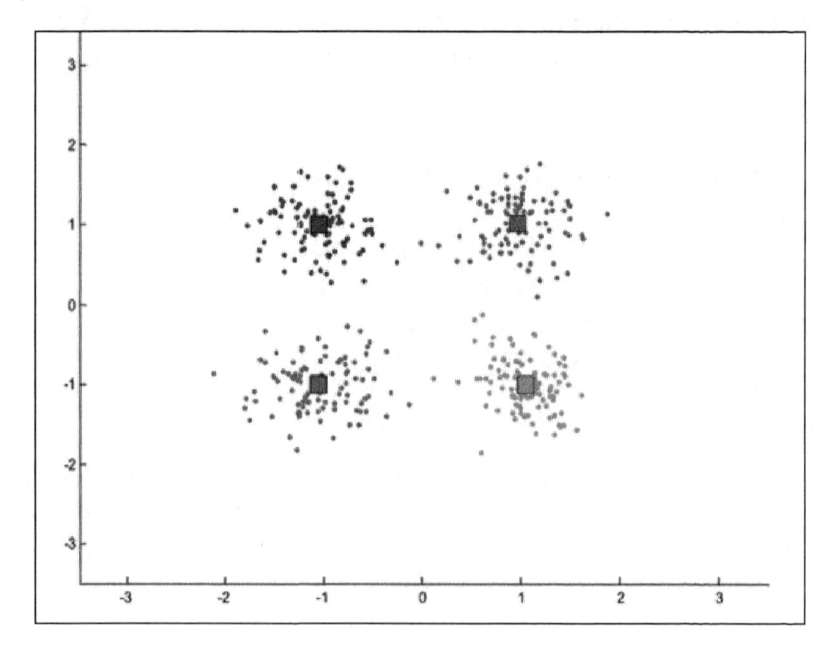

The preceding screenshot is a representation of a cluster analysis. The model will recognize that each uniquely colored cluster of observations is similar to another but different from the other clusters.

A big advantage for unsupervised learning is that it does not require labeled data, which means that it is much easier to get data that complies with unsupervised learning models. Of course, a drawback to this is that we lose all predictive power because the response variable holds the information to make predictions and without it our model will be hopeless in making any sort of predictions.

A big drawback is that it is difficult to see how well we are doing. In a regression or classification problem, we can easily tell how well our models are predicting by comparing our models' answers to the actual answers. For example, if our supervised model predicts rain and it is sunny outside, the model was incorrect. If our supervised model predicts the price will go up by 1 dollar and it goes up by 99 cents, our model was very close! In supervised modeling, this concept is foreign because we have no answer to compare our models to. Unsupervised models are merely suggesting differences and similarities, which then requires a human's interpretation.

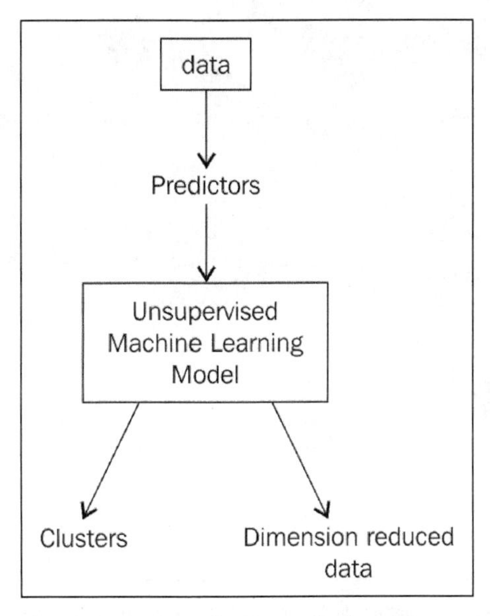

An overview of unsupervised models

In short, the main goal of unsupervised models is to find similarities and differences between data observations. We will discuss unsupervised models in depth in later chapters.

Reinforcement learning

In reinforcement learning, algorithms get to *choose* an action in an environment and then are rewarded (positively or negatively) for choosing this action. The algorithm then adjusts itself and modifies its strategy in order to accomplish some goal, which is usually to get more rewards.

This type of machine learning is very popular in AI-assisted game play as *agents* (the AI) are allowed to explore a virtual world and collect rewards and learn the best navigation techniques. This model is also popular in robotics, especially in the field of self-automated machinery, including cars:

Self-driving cars read in sensor input, act accordingly and are then rewarded for taking a certain action. The car then adjusts its behavior to collect more rewards.

Image source: https://www.quora.com/How-do-Googles-self-driving-cars-work

It can be thought that reinforcement is similar to supervised learning in that the agent is learning from its past actions to make better moves in the future; however, the main difference lies in the reward. The reward does not have to be tied in any way to a "correct" or "incorrect" decision. The reward simply encourages (or discourages) different actions.

Reinforcement learning is the least explored of the three types of machine learning and therefore is not explored in great length in this text. The remainder of the chapter will focus on supervised and unsupervised learning.

Overview of the types of machine learning

Of the three types of machine learning—supervised, unsupervised, and reinforcement learning—we can imagine the world of machine learning as something like this:

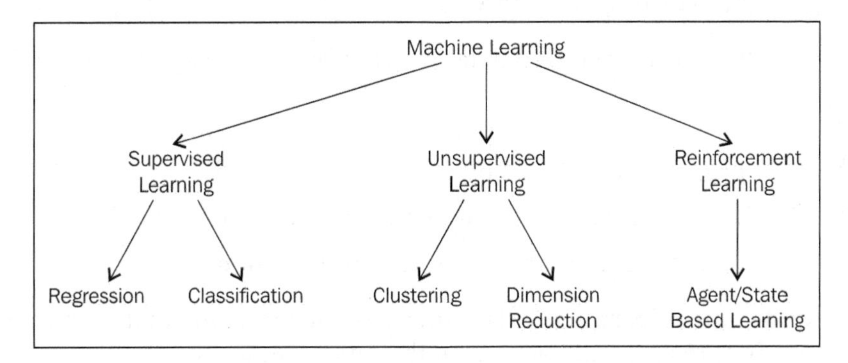

Each of the three types of machine learning has its benefits and also its drawbacks, as listed:

- **Supervised machine learning**: This exploits relationships between predictors and response variables to make predictions of future data observations.

  Pros:

  - It can make future predictions
  - It can quantify relationships between predictors and response variables
  - It can show us how variables affect each other and how much

  Cons:

  - It requires labeled data (which can be difficult to get)

- **Unsupervised machine learning**: This finds similarities and differences between data points.

  Pros:

  - It can find groups of data points that behave similarly that a human would never have noted
  - It can be a preprocessing step for supervised learning

    Think of clustering a bunch of data points and then using these clusters as the response!

  - It can use unlabeled data, which is much easier to find

  Cons:

  - It has zero predictive power
  - It can be hard to determine if we are on the right track
  - It relies much more on human interpretation

- **Reinforcement learning**: This is reward-based learning that encourages agents to take particular actions in their environments.

  Pros:

  - Very complicated rewards systems create very complicated AI systems
  - It can learn in almost any environment, including our own Earth

  Cons:

  - The agent is erratic at first and makes many terrible choices before realizing that these choices have negative rewards

    For example, a car might crash into a wall and not know that that is not okay until the environment negatively rewards it

  - It can take a while before the agent avoids decisions altogether
  - The agent might play it safe and only choose one action and be "too afraid" to try anything else for fear of being punished

How does statistical modeling fit into all of this?

Up until now, I have been using the term machine learning, but you may ask how statistical modeling plays a role in all of this.

This is still a debated topic in the field of data science. I believe that statistical modeling is another term for machine learning models that heavily relies on using mathematical rules borrowed from probability and statistics to create relationships between data variables (often in a predictive sense).

The remainder of this chapter will focus mostly on one statistical/probabilistic model—linear regression.

Linear regression

Finally! We will explore our first true machine learning model. Linear regressions are a form of regression, which means that it is a machine learning model that attempts to find a relationship between predictors and a response variable and that response variable is, you guessed it, continuous! This notion is synonymous with making a *line of best fit*.

In the case of linear regression, we will attempt to find a linear relationship between our predictors and our response variable. Formally, we wish to solve for a formula of the following format:

$$y = \beta_0 + \beta_1 x_1 + \beta_2 x_2 + \cdots + \beta_n x_n$$

- y is our response variable
- x_i is our ith variable (i^{th} column or i^{th} predictor)
- B_0 is the intercept
- B_i is the coefficient for the xi term

Let's take a look at some data before we go in-depth. This dataset is publically available and attempts to predict the number of bikes needed on a particular day for a bike sharing program:

```
# read the data and set the datetime as the index
# taken from Kaggle: https://www.kaggle.com/c/bike-sharing-demand/data
import pandas as pd
import matplotlib.pyplot as plt
%matplotlib inline
```

```
url = 'https://raw.githubusercontent.com/justmarkham/DAT8/master/data/
bikeshare.csv'
bikes = pd.read_csv(url)

bikes.head()
```

	datetime	season	holiday	workingday	weather	temp	atemp	humidity	windspeed	casual	registered	count
0	2011-01-01 00:00:00	1	0	0	1	9.84	14.395	81	0	3	13	16
1	2011-01-01 01:00:00	1	0	0	1	9.02	13.635	80	0	8	32	40
2	2011-01-01 02:00:00	1	0	0	1	9.02	13.635	80	0	5	27	32
3	2011-01-01 03:00:00	1	0	0	1	9.84	14.395	75	0	3	10	13
4	2011-01-01 04:00:00	1	0	0	1	9.84	14.395	75	0	0	1	1

We can see that every row represents a single hour of bike usage. In this case, we are interested in predicting `count`, which represents the total number of bikes rented in the period of that hour.

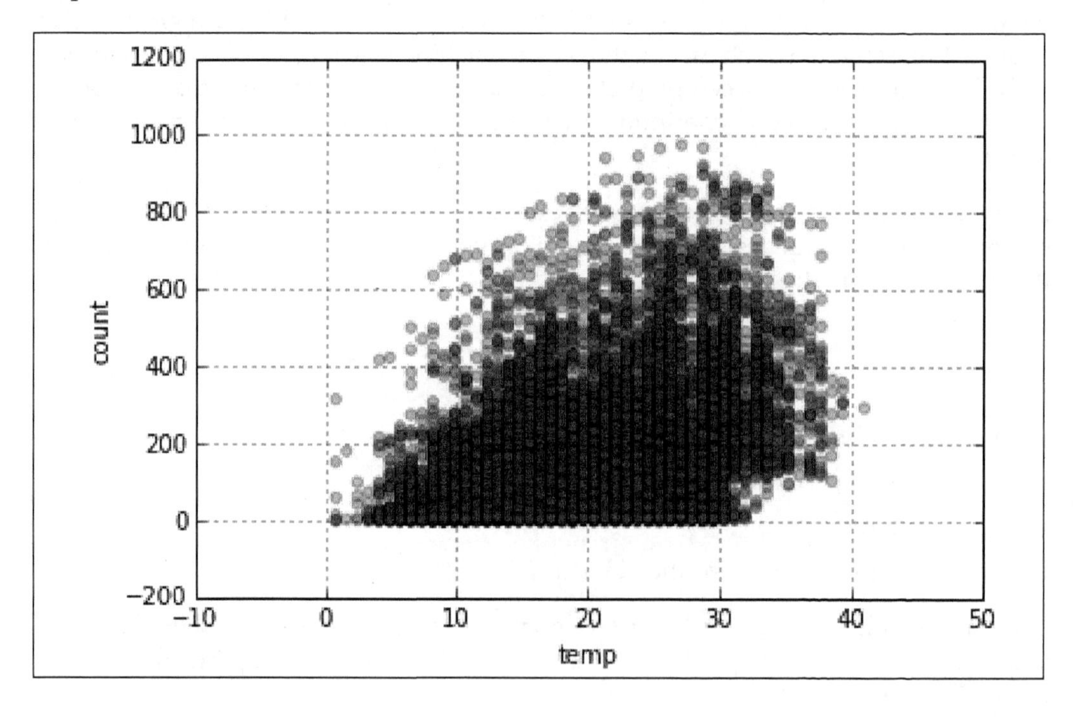

Let's, for example, look at a scatter plot between temperature (the `temp` column) and count, as shown:

```
bikes.plot(kind='scatter', x='temp', y='count', alpha=0.2)
```

And now, let's use a module, called `seaborn`, to draw ourselves a line of best fit, as follows:

```
import seaborn as sns #using seaborn to get a line of best fit
sns.lmplot(x='temp', y='count', data=bikes, aspect=1.5, scatter_
kws={'alpha':0.2})
```

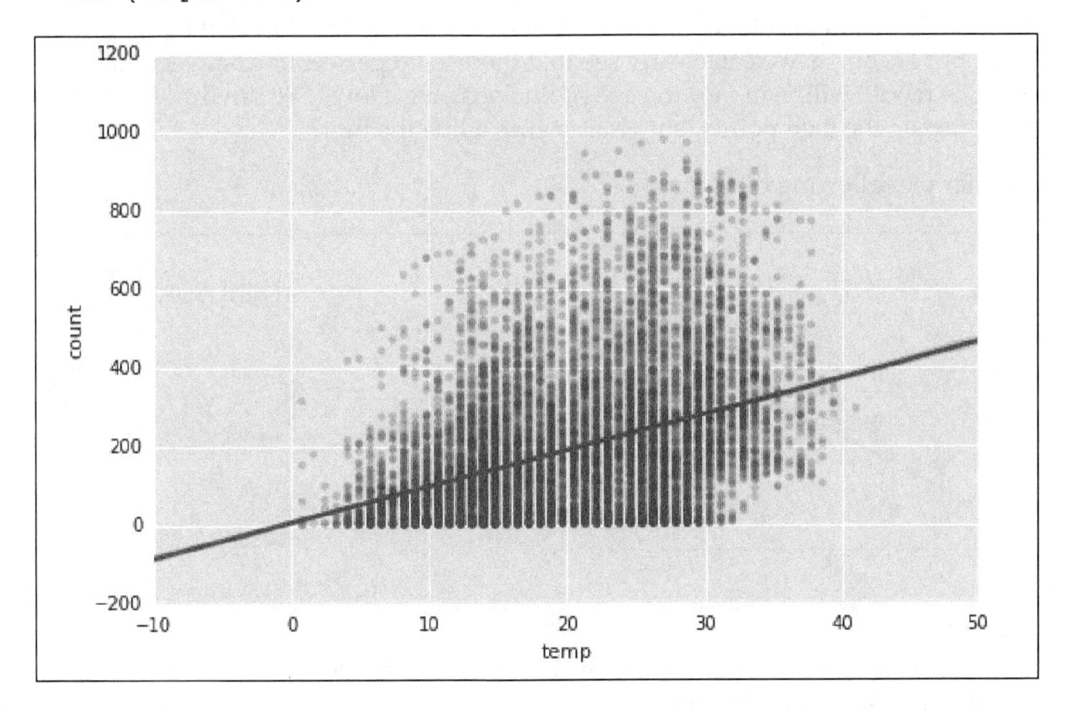

This line in the graph attempts to visualize and quantify the relationship between `temp` and `count`. To make a prediction, we simply find a given temperature, and then see where the line would predict the count. For example, if the temperature is 20 degrees (Celsius mind you), then our line would predict that about 200 bikes will be rented. If the temperature is above 40 degrees, then more than 400 bikes will be needed!

It appears that as `temp` goes up, our `count` also goes up. Let's see if our correlation value, which quantifies a linear relationship between variables, also matches this notion:

```
bikes[['count', 'temp']].corr()
# 0.3944
```

There is a (weak) positive correlation between the two variables! Now, let's go back to the form of the linear regression:

$$y = \beta_0 + \beta_1 x_1 + \beta_2 x_2 + \cdots + \beta_n x_n$$

Our model will attempt to draw a perfect line between all the dots in the preceding graph, but of course, we can clearly see that there is no perfect line between these dots! The model will then find the *best fit* line possible. How? We can draw infinite lines between the data points, but what makes a line the best?

Consider the following diagram:

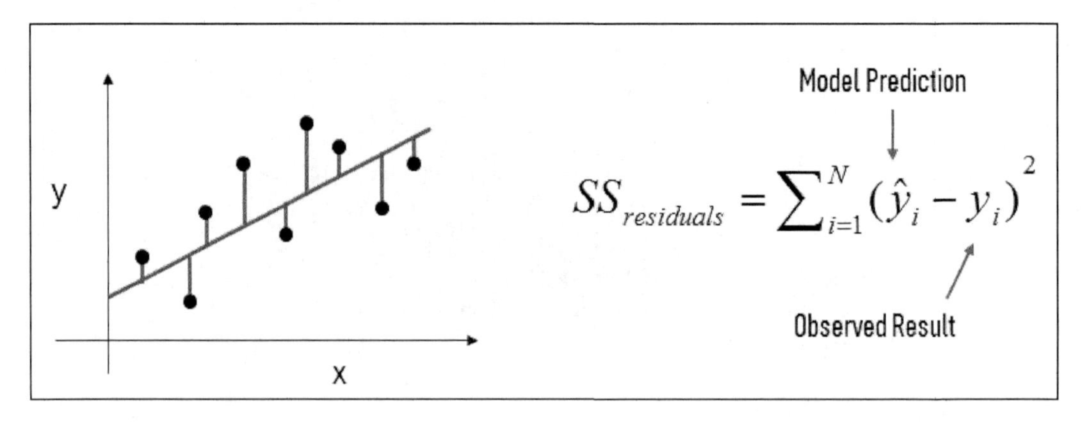

In our model, we are given the *x* and the *y* and the model *learns* the Beta coefficients, also known as **model coefficients**:

- The black dots are the observed values of *x* and *y*.
- The blue line is our line of best fit.
- The red lines between the dots and the line are called the residuals; they are the distances between the observed values and the line. They are how wrong the line is.

Each data point has a residual, or a distance to the line of best fit. The **sum of squared residuals** is the summation of each residual squared. The best fit line has the smallest sum of squared residual value. Let's build this line in Python, shall we?

```
# create X and y
feature_cols = ['temp'] # a lsit of the predictors
X = bikes[feature_cols] # subsetting our data to only the predictors
y = bikes['count'] # our response variable
```

Note how we made an `X` and a `y` variable. These represent our predictors and our response variable.

Then, we will import our machine learning module, `scikit learn`, as shown:

```
# import scikit-learn, our machine learning module
from sklearn.linear_model import LinearRegression
```

Finally, we will fit our model to the predictors and the response variable, as follows:

```
linreg = LinearRegression() #instantiate a new model
linreg.fit(X, y) #fit the model to our data

# print the coefficients
print linreg.intercept_
print linreg.coef_
6.04621295962  # our Beta_0
[ 9.17054048]      # our beta parameters
```

To interpret:

- B_0 *(6.04)* is the value of `y` when `X = 0`.

  It is the estimation of bikes that will be rented when the temperature is 0 Celsius.

  So, at 0 degrees, six bikes are predicted to be in use (its cold!).

Sometimes, it might not make sense to interpret the intercept at all because there might not be a concept of zero of something. Recall the levels of data. Not all levels have this notion. Temperature exists at a level that has the inherent notion of *no bikes;* so, we are safe. Be careful in the future though and always ask yourself, does it make sense to have none of this thing:

- B_1 *(9.17)* is our temp coefficient.
  - It is the change in y divided by the change in x_1.
  - It represents how x and y move together.
  - A change in 1 degree Celsius is associated with an increase of about 9 bikes rented.

- The sign of this coefficient is important. If it were negative, that would imply that a rise in temperature is associated with a drop in rentals.

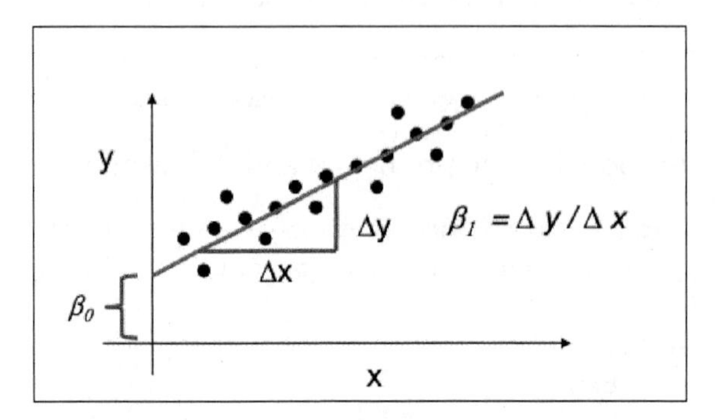

Consider the preceding representation of the Beta coefficients in a linear regression:

It is important to reiterate that these are all statements of correlation and not a statement of causation. We have no means of stating whether or not the rental increase is caused by the change in temperature, it is just that there appears to be movement together.

Using scikit-learn to make predictions is easy!

```
linreg.predict(20)
# 189.4570
```

This means that 190 bikes will likely be rented if the temperature is 20 degrees.

Adding more predictors

Adding more predictors to the model is as simple as telling the linear regression model in scikit-learn about them!

Before we do, we should look at the data dictionary provided to us to make more sense out of these predictors:

- `season`: 1 = spring, 2 = summer, 3 = fall, 4 = winter
- `holiday`: Whether the day is considered a holiday
- `workingday`: Whether the day is a weekend or holiday
- `weather`:
    1. `Clear, Few clouds, Partly cloudy`
    2. `Mist + Cloudy, Mist + Broken clouds, Mist + Few clouds, Mist`

3. `Light Snow, Light Rain + Thunderstorm + Scattered clouds, Light Rain + Scattered clouds`
4. `Heavy Rain + Ice Pallets + Thunderstorm + Mist, Snow + Fog`

- `temp`: Temperature in Celsius
- `atemp`: Feels like temperature in Celsius
- `humidity`: Relative humidity
- `windspeed`: Wind speed
- `casual`: Number of non-registered user rentals initiated
- `registered`: Number of registered user rentals initiated
- `count`: Number of total rentals

Now let's actually create our linear regression model. As before we will first create a list holding the features we wish to look at, create our features and our response datasets (X and y) and then fit our linear regression. Once we fit our regression model, we will take a look at the model's coefficients in order to see how our features are interacting with our response:

```
# create a list of features
feature_cols = ['temp', 'season', 'weather', 'humidity']
# create X and y
X = bikes[feature_cols]
y = bikes['count']

# instantiate and fit
linreg = LinearRegression()
linreg.fit(X, y)

# pair the feature names with the coefficients
zip(feature_cols, linreg.coef_)
```

This gives us:

```
[('temp', 7.8648249924774403),
 ('season', 22.538757532466754),
 ('weather', 6.6703020359238048),
 ('humidity', -3.1188733823964974)]
```

Meaning:

- Holding all other predictors constant, a 1 unit increase in temperature is associated with a rental increase of 7.86 bikes
- Holding all other predictors constant, a 1 unit increase in season is associated with a rental increase of 22.5 bikes

- Holding all other predictors constant, a 1 unit increase in weather is associated with a rental increase of 6.67 bikes
- Holding all other predictors constant, a 1 unit increase in humidity is associated with a rental decrease of 3.12 bikes

This is interesting. Note that as `weather` goes up (meaning that the weather is getting closer to overcast), the bike demand goes up, as is the case when the season variables increase (meaning that we are approaching winter). This is not what I was expecting at all!

Let's take a look at the individual scatter plots between each predictor and the response, as illustrated:

```
feature_cols = ['temp', 'season', 'weather', 'humidity']
# multiple scatter plots
sns.pairplot(bikes, x_vars=feature_cols, y_vars='count', kind='reg')
```

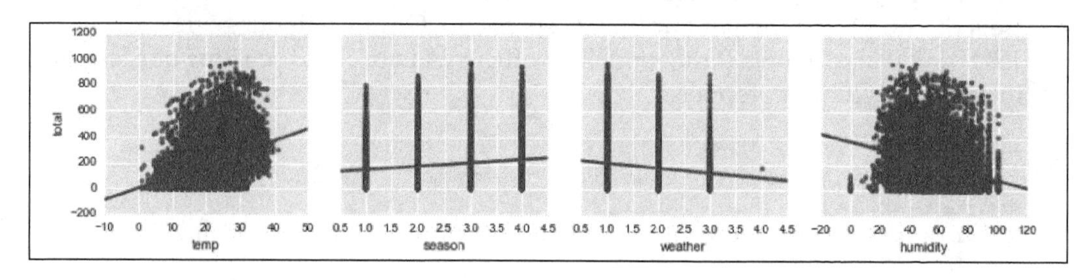

Note how the weather line is trending downwards, which is the opposite of what the last linear model was suggesting. Now, we have to worry about which of these predictors are actually helping us make the prediction, and which ones are just noise. To do so, we're going to need some more advanced metrics.

Regression metrics

There are three main metrics when using regression machine learning models. They are as follows:

- The mean absolute error
- The mean squared error
- The root mean squared error

Each metric attempts to describe and quantify the effectiveness of a regression model by comparing a list of predictions to a list of correct answers. Each of the mentioned metrics is slightly different from the rest and tells a different story.

Mean Absolute Error (MAE) is the mean of the absolute value of the errors:

$$\frac{1}{n}\sum_{i=1}^{n}|y_i - \hat{y}_i|$$

Mean Squared Error (MSE) is the mean of the squared errors:

$$\frac{1}{n}\sum_{i=1}^{n}(y_i - \hat{y}_i)^2$$

Root Mean Squared Error (RMSE) is the square root of the mean of the squared errors:

$$\sqrt{\frac{1}{n}\sum_{i=1}^{n}(y_i - \hat{y}_i)^2}$$

Where:

- n is the number of observations
- y_i is the actual value
- $\hat{y}_i$ is the predicted value

Let's take a look in Python:

```
# example true and predicted response values
true = [9, 6, 7, 6]
pred = [8, 7, 7, 12]
# note that each value in the last represents a single prediction for
a model
# So we are comparing four predictions to four actual answers

# calculate these metrics by hand!
from sklearn import metrics
import numpy as np
print 'MAE:', metrics.mean_absolute_error(true, pred)
print 'MSE:', metrics.mean_squared_error(true, pred)
print 'RMSE:', np.sqrt(metrics.mean_squared_error(true, pred))


MAE: 2.0
MSE: 9.5
RMSE: 3.08220700148
```

The breakdown of these numbers is as follows:

- **MAE** is probably the easiest to understand, because it's just the average error. It denotes, on an average, how wrong the model is.
- **MSE** is more effective than MAE, because MSE punishes larger errors, which tends to be much more useful in the real world.
- **RMSE** is even more popular than MSE, because it is much more interpretable.

RMSE is usually the preferred metric for regression, but no matter which one you choose, they are all loss functions and therefore are something to be minimized. Let's use the RMSE to ascertain which columns are helping and which are hurting.

Let's start with only using temperature. Note that our procedure will be as follows:

1. Create our `X` and our `y` variables.
2. Fit a linear regression model.
3. Use the model to make a list of predictions based on `X`.
4. Calculate the RMSE between the predictions and the actual values.

Let's take a look at the code:

```
from sklearn import metrics
# import metrics from scikit learn

feature_cols = ['temp']
# create X and y
X = bikes[feature_cols]
linreg = LinearRegression()
linreg.fit(X, y)
y_pred = linreg.predict(X)
np.sqrt(metrics.mean_squared_error(y, y_pred)) # RMSE
# Can be interpreted loosely as an average error
#166.45
```

Now, let's try it using temperature and humidity, as shown:

```
feature_cols = ['temp', 'humidity']
# create X and y
X = bikes[feature_cols]
linreg = LinearRegression()
linreg.fit(X, y)
y_pred = linreg.predict(X)
np.sqrt(metrics.mean_squared_error(y, y_pred)) # RMSE
# 157.79
```

It got better!! Let's try using even more predictors, as illustrated:

```
feature_cols = ['temp', 'humidity', 'season', 'holiday', 'workingday',
'windspeed', 'atemp']
# create X and y
X = bikes[feature_cols]
linreg = LinearRegression()
linreg.fit(X, y)
y_pred = linreg.predict(X)
np.sqrt(metrics.mean_squared_error(y, y_pred)) # RMSE
# 155.75
```

Even better! At first, this seems like a major triumph, but there is actually a hidden danger here. Note that we are training the line to fit to X and y and, then, asking it to predict X again! This is actually a huge mistake in machine learning because it can lead to **overfitting**, which means that our model is merely *memorizing* the data and regurgitating it back to us.

Imagine that you are a student, and you walk into the first day of class and the teacher says that the final exam is very difficult in this class. In order to prepare you, she gives you practice test after practice test after practice test. The day of the final exam arrives and you are shocked to find out that every question on the exam is exactly the same as in the practice test! Luckily, you did them so many times that you remember the answer and get a 100% in the exam.

The same thing applies here, more or less. By fitting and predicting on the same data, the model is memorizing the data and getting better at it. A great way to combat this overfitting problem is to use the train/test approach to fit machine learning models, which works as illustrated:

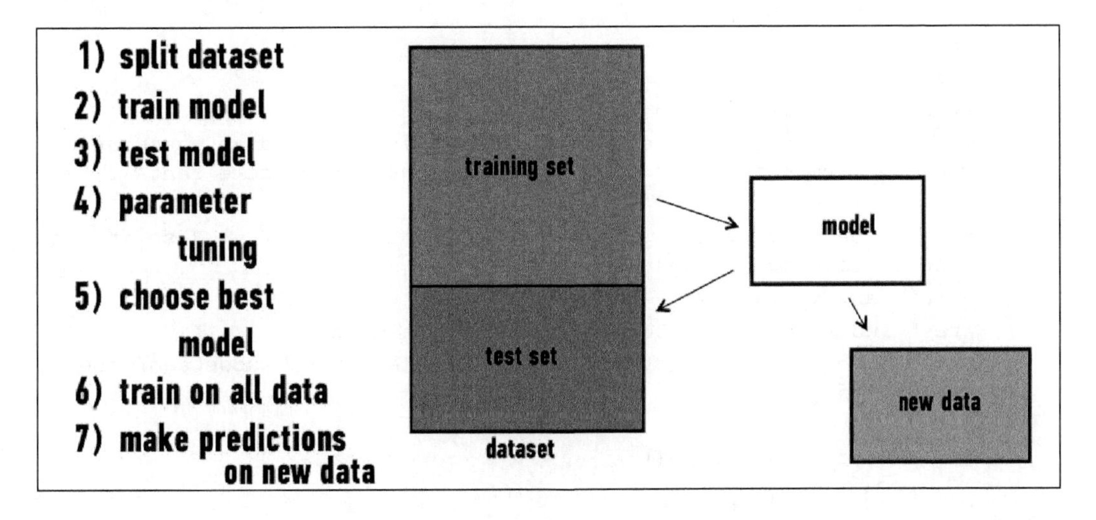

Essentially, we will take the following steps:

1. Split up the dataset into two parts: a training and a test set.
2. Fit our model on the training set and then test it on the testing set. Just like in school, where the teacher would teach from one set of notes and then test us on different (but similar) questions.
3. Once our model is good enough (based on our metrics), we turn our model's attention toward the entire dataset.
4. Our model awaits for new data previously unseen by anyone.

The goal here is to minimize the out-of-sample errors of our model, which are the errors our model has on data that it has never seen before. This is important because the main idea (usually) of a supervised model is to predict unseen test cases. If our model is unable to generalize from our training data and use that to predict unseen cases, then our model isn't very good.

The preceding diagram outlines a simple way of ensuring that our model can effectively ingest the training data and use it to predict data points that the model itself has never seen. Of course, as data scientists, we know that the test set also has answers attached to them, but the model doesn't know that.

All of this might sound complicated, but luckily, scikit-learn has a built-in method to do this, as shown:

```
from sklearn.cross_validation import train_test_split
# function that splits data into training and testing sets


feature_cols = ['temp']
X = bikes[feature_cols]
y = bikes['count']
# setting our overall data X, and y
# Note that in this example, we are attempting to find an association
between the temperature of the day and the number of bike rentals.

X_train, X_test, y_train, y_test = train_test_split(X, y) # split the
data into training and testing sets
# X_train and y_train will be used to train the model
# X_test and y_test will be used to test the model
# Remember that all four of these variables are just subsets of the
overall X and y.

linreg = LinearRegression()
```

```
# instantiate the model

linreg.fit(X_train, y_train)
# fit the model to our training set

y_pred = linreg.predict(X_test)
# predict our testing set

np.sqrt(metrics.mean_squared_error(y_test, y_pred)) # RMSE
# Calculate our metric: 166.91
```

We will spend more time on the reasoning behind this train/test split in *Chapter 12, Beyond the Essentials,* and look into an even more helpful method, but the main reason we must go through this extra work is because we do not want to fall into a trap where our model is simply regurgitating our dataset back to us and will not be able to handle unseen data points.

In other words, our train test split is ensuring that the metrics we are looking at are more honest estimates of our sample performance.

Now, let's try again with more predictors, as follows:

```
feature_cols = ['temp', 'workingday']
X = bikes[feature_cols]
y = bikes['count']

X_train, X_test, y_train, y_test = train_test_split(X, y)
# Pick a new random training and test set

linreg = LinearRegression()
linreg.fit(X_train, y_train)
y_pred = linreg.predict(X_test)
# fit and predict

np.sqrt(metrics.mean_squared_error(y_test, y_pred))
# 166.95
```

Now our model actually got worse with that addition! Implying that `workingday` might not be very predictive of our response, the bike rental count.

Now, all of this is good and well, but how well is our model really doing at predicting? We have our root mean squared error of around 167 bikes, but is that *good*? One way to discover this is to evaluate the **null model**.

The null model in supervised machine learning represents effectively guessing the expected outcome over and over, and seeing how you did. For example, in regression, if we only ever guess the average number of hourly bike rentals, then how well would that model do?

First, let's get the average hourly bike rental, as shown:

```
average_bike_rental = bikes['count'].mean()
average_bike_rental
# 191.57
```

This means that, overall, in this dataset, regardless of weather, time, day of the week, humidity, and everything else, the average number of bikes that go out every hour is about 192.

Let's make a *fake* prediction list, wherein every single guess is 191.57. Let's make this guess for every single hour, as follows:

```
num_rows = bikes.shape[0]
num_rows
# 10886
All 10,886 of them.
null_model_predictions = [average_bike_rental]*num_rows
null_model_predictions
[191.57413191254824,
 191.57413191254824,
 191.57413191254824,
 191.57413191254824,
...
 191.57413191254824,
 191.57413191254824,
 191.57413191254824,
 191.57413191254824]
```

So, now we have 10,886 values, all of them are the average hourly bike rental number. Now, let's see what the root mean squared error would be if our model only ever guessed the expected value of the average hourly bike rental count:

```
np.sqrt(metrics.mean_squared_error(y, null_model_predictions))
181.13613
```

Simply guessing, it looks like our root mean squared error would be 181 bikes. So, even with one or two features, we can beat it! Beating the null model is a kind of baseline in machine learning. If you think about it, why go through any effort at all if your machine learning is not even better than just guessing!

We've spent a great deal of time on linear regression, but I'd like to now take some time to look at our next machine learning model, which is actually, somewhat, a cousin of linear regression. They are based on very similar ideas but have one major difference— while linear regression is a regression model and can only be used to make predictions of continuous numbers, our next machine learning model will be a classification model, which means that it will attempt to make associations between features and a categorical response.

Logistic regression

Our first classification model is called logistic regression. I can already hear the questions you have in your head: what makes is logistic, why is it called regression if you claim that this is a classification algorithm? All in good time, my friend.

Logistic regression is a generalization of the linear regression model adapted to fit classification problems. In linear regression, we use a set of quantitative feature variables to predict a continuous response variable. In logistic regression, we use a set of quantitative feature variables to predict *probabilities* of class membership. These probabilities can then be mapped to class labels, thus predicting a class for each observation.

When performing linear regression, we use the following function to make our line of best fit:

$$y=\beta_0+\beta_1 x$$

Here, y is our response variable (the thing we wish to predict), our Beta represents our model parameters and x represents our input variable (a single one in this case, but it can take in more, as we have seen).

Briefly, let's assume that one of the response options in our classification problem is the class 1.

When performing logistic regression, we use the following form:

$$\pi=\Pr(y=1\mid x)=\frac{e^{\beta_0+\beta_1 x}}{1+e^{\beta_0+\beta_1 x}}$$

Probability of y = 1, given x

Here, π represents the conditional probability that our response variable belongs to class 1, given our data *x*. Now, you might be wondering what on earth is that monstrosity of a function on the right-hand side, and where did the *e* variable come from? Well, that monstrosity is called the logistic function, and it is actually wonderful. And that variable, *e*, is no variable at all. Let's back up a tick.

The variable *e* is a special number, like π . It is, approximately, 2.718, and is called **Euler's number**. It is used frequently in modeling environments with *natural* growth and decay. For example, scientists use e in order to model the population growth of bacteria and buffalo alike. Euler's number is used to model the radioactive decay of chemicals and also to calculate continuous compound interest! Today, we will use e for a very special purpose, for machine learning.

Why can't we just make a linear regression directly to the probability of the data point belonging to a certain class like this?

$$\Pr(y=1|x)=y=\beta_0+\beta_1 x$$

We can't do that for a few reasons, but I will point out a big one. Linear regression, because it attempts to relate to a continuous response variable, assumes that our y is continuous. In our case, y would represent the probability of an event occurring. Even though our probability is, in fact, a continuous range, it is just that—a range between 0 and 1. A line would extrapolate beyond 0 and 1 and *be able* to predict a probability of -4 or 1,542! We can't have that. Our graph must be bound neatly between 0 and 1 on the y axis like a real probability is.

Another reason is a bit more philosophical. Using a linear regression, we are making a serious assumption. Our big assumption here is that there is a linear relationship between probability and our features. In general, if we think about the probability of an event, we tend to think of smooth curves representing them, not a single boring line. So, we need something a bit more appropriate. For this, let us go back and revisit basic probability for a minute.

Probability, odds, and log odds

We are familiar with the basic concept of probability in that the probability of an event occurring can be simply modeled as the number of ways the event can occur divided by all the possible outcomes. For example, if, out of *3,000* people who walked into a store, *1,000* actually bought something, then we could say that the probability of a single person buying an item is as shown:

$$\Pr(buy) = \frac{1,000}{3,000} = \frac{1}{3} = 33.3\%$$

However, we also have a related concept, called **odds**. The odds of an outcome occurring is the ratio of the number of ways that the outcome occurs divided by every other possible outcome instead of all possible outcomes. In the same example, the odds of a person buying something would be as follows:

$$Odds(buy) = \frac{1,000}{3,000} = \frac{1}{2} = .5$$

This means that for every customer you convert, you will *not* convert two customers. These concepts are so related, there is even a formula to get from one to the other. We have that:

$$Odds = \frac{P}{1-P}$$

Let's check this with our example, as illustrated:

$$Odds = \frac{\frac{1}{3}}{1-\frac{1}{3}} = \frac{\frac{1}{3}}{\frac{2}{3}} = \frac{1}{2}$$

It checks out!

Let's use Python to make a table of probabilities and associated odds, as shown:

```
# create a table of probability versus odds
table = pd.DataFrame({'probability':[0.1, 0.2, 0.25, 0.5, 0.6, 0.8,
0.9]})
table['odds'] = table.probability/(1 - table.probability)
table
```

	probability	odds
0	0.10	0.111111
1	0.20	0.250000
2	0.25	0.333333
3	0.50	1.000000
4	0.60	1.500000
5	0.80	4.000000
6	0.90	9.000000

So, we see that as our probabilities increase, so do our odds, but at a much faster rate! In fact, as the probability of an event occurring nears 1, our odds will shoot off into infinity. Earlier, we said that we couldn't simply regress to probability because our line would shoot off into positive and negative infinities, predicting improper probabilities, but what if we regress to odds? Well, odds go off to positive infinity, but alas, they will merely approach 0 on the bottom, but never go below 0. Therefore, we cannot simply regress to probability, or odds. It looks like we've hit rock bottom folks!

However, wait, natural numbers and logarithms to the rescue! Think of logarithms as follows:

$$if\, 2^4 = 16 \; then \; \log_2 16 = 4$$

Basically, logarithms and exponents are one and the same. We are just so used to writing exponents in the first way that we forget there is another way to write them. How about another example? If we take the logarithm of a number, we are asking the question, *hey, what exponent would we need to put on this number to make it the given number?*

Note that `np.log` automatically does all logarithms in base e, which is what we want:

```
np.log(10) # == 2.3025
# meaning that e ^ 2.302 == 10

# to prove that
2.71828**2.3025850929940459 # == 9.9999
# e ^ log(10) == 10
```

Let's go ahead and add the logarithm of odds, or log-odds to our table, as follows:

```
# add log-odds to the table
table['logodds'] = np.log(table.odds)
table
```

	probability	odds	logodds
0	0.10	0.111111	-2.197225
1	0.20	0.250000	-1.386294
2	0.25	0.333333	-1.098612
3	0.50	1.000000	0.000000
4	0.60	1.500000	0.405465
5	0.80	4.000000	1.386294
6	0.90	9.000000	2.197225

So, now every row has the probability of a single event occurring, the odds of that event occurring, and now the log-odds of that event occurring. Let's go ahead and ensure that our numbers are on the up and up. Let's choose a probability of `.25`, as illustrated:

```
prob = .25

odds = prob / (1 - prob)
odds
# 0.33333333

logodds = np.log(odds)
logodds
# -1.09861228
```

It checks out! Wait, look! Our `logodds` variable seems to go down below zero and, in fact, `logodds` is not bounded above, nor is it bounded below, which means that it is a great candidate for a response variable for linear regression. In fact, this is where our story of logistic regression really begins.

The math of logistic regression

The long and short of it is that logistic regression is a linear regression between our feature, X, and the log-odds of our data belonging to a certain class that we will call *true* for the sake of generalization.

If p represents the probability of a data point belonging to a particular class, then logistic regression can be written as follows:

$$\log_e\left(\frac{p}{1-p}\right)=\beta_0+\beta_1 x$$

If we rearrange our variables and solve this for p, we would get the logistic function, which takes on an S shape, where y is bounded by `[0,1]`:

$$p=\frac{e^{\beta_0+\beta_1 x}}{1+e^{\beta_0+\beta_1 x}}$$

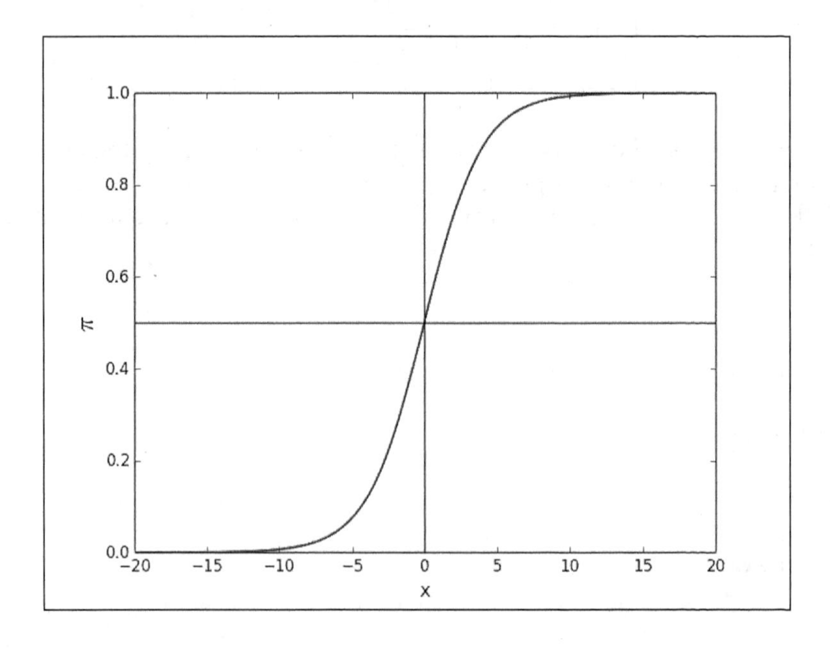

The preceding graph represents the logistic function's ability to map our continuous input, x, to a smooth probability curve that begins at the left, near probability 0, and as we increase x, our probability of belonging to a certain class rises *naturally* and smoothly up to probability 1. In other words:

- Logistic regression gives an output of the probabilities of a specific class being true
- Those probabilities can be converted into class predictions

The logistic function has some nice properties, as follows:

It takes on an *S* shape

- Output is bounded by 0 and 1, as a probability should be

In order to interpret the outputs of a logistic function, we must understand the difference between probability and odds. The odds of an event are given by the ratio of the probability of the event by its complement, as shown:

$$odds = \frac{p}{1-p}$$

In linear regression, the β_1 parameter represents the change in the response variable for a unit change in x. In logistic regression, β_1 represents the change in the log-odds for a unit change in x. This means that e^{β_1} gives us the change in the odds for a unit change in x.

Consider that we are interested in mobile purchase behavior. Let y be a class label denoting purchase/no purchase, and let x denote whether the phone was an iPhone.

Also, suppose that we perform a logistic regression, and we get β_1 = 0.693.

In this case the odds ratio is *np.exp(0.693)* = 2, which means that the likelihood of purchase is twice as high if the phone is an iPhone.

Our examples have mostly been binary classification, meaning that we are only predicting one of two outcomes but logistic regression can handle predicting multiple options in our categorical response using a one-versus-all approach, meaning that it will fit a probability curve for each categorical response!

Back to our bikes briefly to see scikit-learn's logistic regression in action. I will begin by making a new response variable that is categorical. To make things simple, I made a column, called `above_average`, which is true if the hourly bike rental count is above average and false otherwise.

```
# Make a cateogirical response
bikes['above_average'] = bikes['count'] >= average_bike_rental
```

As mentioned before, we should look at our null model. In regression, our null model always predicts the average response, but in classification, our null model always predicts the most common outcome. In this case, we can use a Pandas value count to see that. About 60% of the time, the bike rental count is not above average:

```
bikes['above_average'].value_counts(normalize=True)
```

Now, let's actually use logistic regression to try and predict whether or not the hourly bike rental count will be above average, as shown:

```
from sklearn.linear_model import LogisticRegression

feature_cols = ['temp']
# using only temperature

X = bikes[feature_cols]
y = bikes['above_average']
# make our overall X and y variables, this time our y is
# out binary response variable, above_average

X_train, X_test, y_train, y_test = train_test_split(X, y)
# make our train test split

logreg = LogisticRegression()
# instantate our model

logreg.fit(X_train, y_train)
# fit our model to our training set

logreg.score(X_test, y_test)
# score it on our test set to get a better sense of out of sample
performance

# 0.65650257
```

It seems that by only using temperature, we can beat the null model of guessing false all of the time! This is our first step in making our model the best it can be.

Between linear and logistic regression, I'd say we already have a great tool belt of machine learning forming, but I have a question—it seems that both of these algorithms are only able to take in quantitative columns as features, but what if I have a categorical feature that I think has an association to my response?

Dummy variables

Dummy variables are used when we are hoping to convert a categorical feature into a quantitative one. Remember that we have two types of categorical features: nominal and ordinal. Ordinal features have natural order among them, while nominal data does not.

Encoding qualitative (nominal) data using separate columns is called making dummy variables and it works by turning each unique category of a nominal column into its own column that is either true or false.

For example, if we had a column for someone's college major and we wished to plug that information into a linear or logistic regression, we couldn't because they only take in numbers! So, for each row, we had new columns that represent the single nominal column. In this case, we have four unique majors: computer science, engineering, business, and literature. We end up with three new columns (we omit computer science as it is not necessary).

Major (k=4)
Computer Science
Engineering
Business
Literature
Business
Engineering

Engineering	Business	Literature
0	0	0
1	0	0
0	1	0
0	0	1
0	1	0
1	0	0

Note that the first row has a 0 in all the columns, which means that this person did *not* major in engineering, did *not* major in business and did *not* major in literature. The second person has a single 1 in the engineering column as that is the major they studied.

In our bikes example, let's define a new column, called `when_is_it`, which is going to be one of the following four options:

- `Morning`
- `Afternoon`
- `Rush_hour`
- `Off_hours`

To do this, our approach will be to make a new column that is simply the hour of the day, use that column to determine when in the day it is, and explore whether or not we think that column might help us predict the `above_daily` column:

```
bikes['hour'] = bikes['datetime'].apply(lambda x:int(x[11]+x[12]))
# make a column that is just the hour of the day
bikes['hour'].head()
0
1
2
3
```

Great, now let's define a function that turns these hours into strings. For this example, let's define the hours between 5 and 11 as morning, between 11 am and 4 pm as being afternoon, 4 and 6 as being rush hours, and everything else as being off hours:

```
# this function takes in an integer hour
# and outputs one of our four options
def when_is_it(hour):
    if hour >= 5 and hour < 11:
        return "morning"
    elif hour >= 11 and hour < 16:
        return "afternoon"
    elif hour >= 16 and hour < 18:
        return "rush_hour"
    else:
        return "off_hours"
```

Let's apply this function to our new `hour` column and make our brand new column, `when_is_it`:

```
bikes['when_is_it'] = bikes['hour'].apply(when_is_it)
bikes[['when_is_it', 'above_average']].head()
```

	when_is_it	above_average
0	off_hours	False
1	off_hours	False
2	off_hours	False
3	off_hours	False
4	off_hours	False

Let's try to use only this new column to determine whether or not the hourly bike rental count will be above average. Before we do, let's do the basics of exploratory data analysis and make a graph to see if we can visualize a difference between the four times of the day. Our graph will be a bar chart with one bar per time of the day. Each bar will represent the percentage of times that this time of the day had a greater than normal bike rental:

```
bikes.groupby('when_is_it').above_average.mean().plot(kind='bar')
```

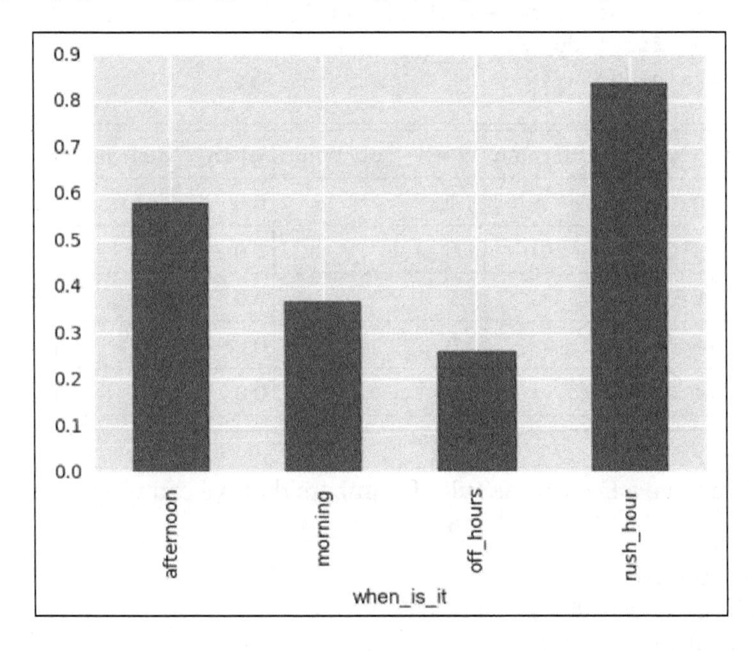

We can see that there is a pretty big difference! For example, when it is off hours, the chance of having more than average bike rentals is about 25%, whereas during rush hours, the chance of being above average is over 80%! Okay, this is exciting, but let's use some built-in Pandas tools to extract dummy columns, as follows:

```
when_dummies = pd.get_dummies(bikes['when_is_it'], prefix='when__')
when_dummies.head()
```

	when__afternoon	when__morning	when__off_hours	when__rush_hour
0	0.0	0.0	1.0	0.0
1	0.0	0.0	1.0	0.0
2	0.0	0.0	1.0	0.0
3	0.0	0.0	1.0	0.0
4	0.0	0.0	1.0	0.0

```
when_dummies = when_dummies.iloc[:, 1:]
# remove the first column
when_dummies.head()
```

	when__morning	when__off_hours	when__rush_hour
0	0.0	1.0	0.0
1	0.0	1.0	0.0
2	0.0	1.0	0.0
3	0.0	1.0	0.0
4	0.0	1.0	0.0

Great! Now we have a Dataframe full of numbers that we can plug in to our logistic regression:

```
X = when_dummies
# our new X is our dummy variables
y = bikes.above_average

logreg = LogisticRegression()
# instantate our model
```

```
logreg.fit(X_train, y_train)
# fit our model to our training set

logreg.score(X_test, y_test)
# score it on our test set to get a better sense of out of sample
performance

# 0.685157
```

Which is even better than just using the temperature! What if we tacked on temperature and humidity onto that? So, now we are using the temperature, humidity, and our time of day dummy variables to predict whether or not we will have higher than average bike rentals:

```
new_bike = pd.concat([bikes[['temp', 'humidity']], when_dummies],
axis=1)
# combine temperature, humidity, and the dummy variables

X = new_bike
# our new X is our dummy variables
y = bikes.above_average

X_train, X_test, y_train, y_test = train_test_split(X, y)

logreg = LogisticRegression()
# instantate our model

logreg.fit(X_train, y_train)
# fit our model to our training set

logreg.score(X_test, y_test)
# score it on our test set to get a better sense of out of sample
performance

# 0.7182218
```

Wow. Okay, let's quit while we're ahead.

Summary

In this chapter, we looked at machine learning and its different subcategories. We explored supervised, unsupervised, and reinforcement learning strategies, and looked at situations where each one would come in handy.

Looking into linear regression, we were able to find relationships between predictors and a continuous response variable. Through the train/test split, we were able to help avoid overfitting our machine learning models and get a more generalized prediction. We were able to use metrics, such as the root mean squared error, to evaluate our models as well.

By extending our notion of linear regression into logistic regression, we were able to then find association between the same predictors, but now to categorical responses.

By introducing dummy variables into the mix, we were able to add categorical features to our models and improve our performance even further.

In the next few chapters, we will be taking a much deeper dive into many more machine learning models and, along the way, we will learn new metrics, new validation techniques, and more importantly, new ways of applying our data science to the world.

11

Predictions Don't Grow on Trees – or Do They?

In this chapter, we will be looking at three types of machine learning algorithms. The first two being examples of supervised learning while the final algorithm being an example of unsupervised learning.

Our goal in this chapter is to see and apply concepts learned from previous chapters in order to construct and use modern learning algorithms in order to glean insights and make predictions on real data sets. While we explore the following algorithms, we should always remember that we are constantly keeping our metrics in mind.

Let's get to it!

Naïve Bayes classification

Let's get right into it! Let's begin with Naïve Bayes classification. This machine learning model relies heavily on results from previous chapters, specifically with Bayes theorem:

$$P(H \mid D) = \frac{P(D \mid H)P(H)}{P(D)}$$

Let's look a little closer at the specific features of this formula:

- *P(H)* is the probability of the hypothesis before we observe the data, called the **prior probability**, or just **prior**
- *P(H | D)* is what we want to compute, the probability of the hypothesis after we observe the data, called the **posterior**

- *P(D | H)* is the probability of the data under the given hypothesis, called the **likelihood**
- *P(D)* is the probability of the data under any hypothesis, called the **normalizing constant**

Naïve Bayes classification is a classification model, and therefore a supervised model. Given this, what kind of data do we need?

- Labeled data
- Unlabeled data

(*Insert jeopardy music here*)

If you answered labeled data then you're well on your way to becoming a data scientist!

Suppose we have a data set with *n* features, (`x1`, `x2`, ..., `xn`) and a class label C. For example let's take some data involving spam text classification. Our data would consist of rows of individual text samples and columns of both our features and our class labels. Our features would be words and phrases that are contained within the text samples and our class labels are simply `spam` or `not spam`. In this scenario, I will replace the class `not spam` with the easier to say word, `ham`:

```
import pandas as pd
import sklearn
df = pd.read_table('https://raw.githubusercontent.com/sinanuozdemir/
sfdat22/master/data/sms.tsv',
                    sep='\t', header=None, names=['label', 'msg'])
df
```

Here is a sample of text data in a row column format:

	label	msg
0	ham	Go until jurong point, crazy.. Available only ...
1	ham	Ok lar... Joking wif u oni...
2	spam	Free entry in 2 a wkly comp to win FA Cup fina...
3	ham	U dun say so early hor... U c already then say...
4	ham	Nah I don't think he goes to usf, he lives aro...
5	spam	FreeMsg Hey there darling it's been 3 week's n...
6	ham	Even my brother is not like to speak with me. ...
7	ham	As per your request 'Melle Melle (Oru Minnamin...

Let's do some preliminary statistics to see what we are dealing with. Let's see the difference in the number of ham and spam messages at our disposal:

```
df.label.value_counts().plot(kind="bar")
```

This gives us a bar chart, as follows:

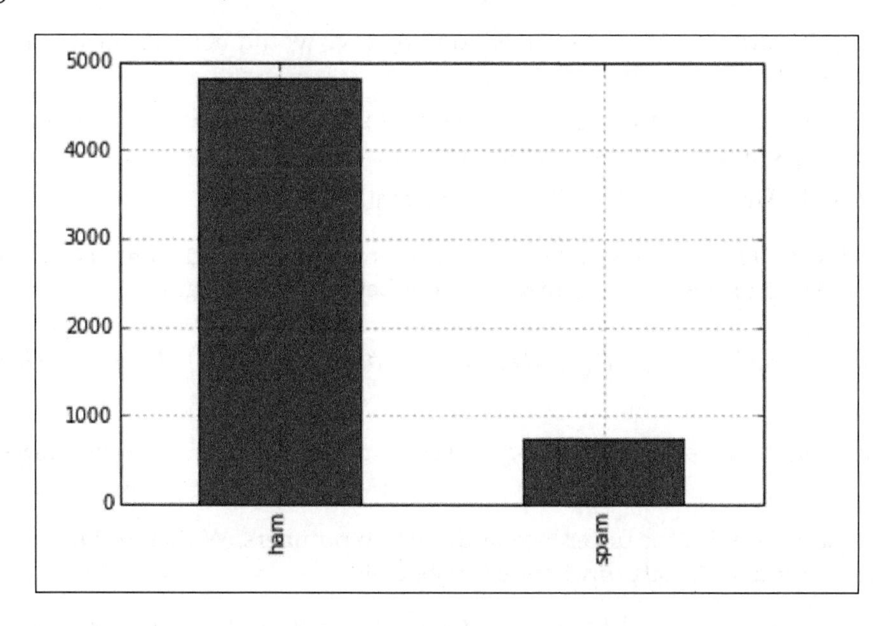

So we have WAY more ham messages than we do spam. Because this is a classification problem, it will be very useful to know our *null accuracy rate* which is the percentage chance of predicting a single row correctly if we keep guessing the most common class, *ham*:

```
df.label.value_counts() / df.shape[0]

ham      0.865937
spam     0.134063
```

So if we blindly guessed `ham` we would be correct about 87% of the time, but we can do better than that. If we have a set of classes, C, and a features *xi*, then we can use Bayes theorem to predict the probability that a single row belongs to class C using the following formula:

$$P\left(class\,C \mid \{x_i\}\right) = \frac{P\left(\{x_i\} \mid class\,C\right) \cdot P\left(class\,C\right)}{P\left(\{x_i\}\right)}$$

Let's look at this formula in a little more detail:

- *P(class C | {xi})*: The posterior probability is the probability that the row belongs to *class* C given the features *{xi}*.
- *P({xi} | class C)*: This is the likelihood that we would observe these features given that the row was in *class* C.
- *P(class C)*: This is the prior probability. It is the probability that the data point belongs to *class* C before we see any data.
- *P({xi})*: This is our normalization constant.

For example, imagine we have an e-mail with three words: `send cash now`. We'll use Naïve Bayes to classify the e-mail as either being spam or ham:

$$P\left(spam \mid send\,cash\,now\right) = P\left(send\,cash\,now \mid spam\right) * P\left(spam\right) / P\left(send\,cash\,now\right)$$

$$P\left(ham \mid send\,cash\,now\right) = P\left(send\,cash\,now \mid ham\right) * P\left(ham\right) / P\left(send\,cash\,now\right)$$

We are concerned with the difference of these two numbers. We can use the following criteria to classify any single text sample:

- If `P(spam | send cash now)` is larger than `P(ham | send cash now)`, then we will classify the text as spam

- If `P(ham | send cash now)` is larger than `P(spam | send cash now)`, then we will label the text as ham

Because both equations have *P (send money now)* on the denominator, we can ignore them.

So now we are concerned with the following:

$$P(send\ cash\ now \mid spam) * P(spam)\ VS\ P(send\ cash\ now \mid ham) * P(ham)$$

Let's figure out the numbers in this equation:

- *P(spam) = 0.134063*
- *P(ham) = 0.865937*
- *P(send cash now | spam)*
- *P(send cash now | ham)*

The final two likelihoods might seem like they would not be so difficult to calculate. All we have to do is count the numbers of spam messages that include the phrase `send money now` and divide that by the total number of spam messages:

```
df.msg = df.msg.apply(lambda x:x.lower())
# make all strings lower case so we can search easier


df[df.msg.str.contains('send cash now')] .shape
(0, 2)
```

Oh no! There are none! There are literally `0` texts with the exact phrase `send cash now`. The hidden problem here is that this phrase is very specific and we can't assume that we will have enough data in the world to have seen this exact phrase many times before. Instead we can make a *naïve assumption* in our Bayes theorem. If we assume that the features (words) are conditionally independent (meaning that no word affects the existence of another word) then we can rewrite the formula:

$$P(send\ cash\ now \mid spam) = P(send \mid spam) * P(cash \mid spam) * P(now \mid spam)$$

```
spams = df[df.label == 'spam']
for word in ['send', 'cash', 'now']:
    print word, spams[spams.msg.str.contains(word)].shape[0] /
float(spams.shape[0])revealing
```

- *P(send | spam) = 0.096*
- *P(cash | spam) = 0.091*
- *P(now | spam) = 0.280*

Meaning we can calculate the following:

$$P(send\ cash\ now \mid spam) * P(spam) = (.096 * .091 * .280) * .134 = 0.00032$$

Repeating the same procedure for ham gives us the following:

- *P(send | ham) = 0.03*
- *P(cash | ham) = 0.003*
- *P(now | ham) = 0.109*

$$P(send\ cash\ now \mid ham) * P(ham) = (.03 * .003 * .109) * .865 = 0.0000084$$

The fact that these numbers are both very low is not as important as the fact that the spam probability is much larger than the ham calculation. If we calculate *.00032 / .0000084 = 38.1* we see that the `send cash now` probability for spam is 38 times higher than for spam.

Doing this means that we can classify `send cash now` as spam! Simple, right?

Let's use Python to implement a Naïve Bayes classifier without having to do all of these calculations ourselves.

First, let's revisit the count vectorizer in scikit-learn that turns text into numerical data for us. Let's assume that we will train on three documents (sentences):

```
# simple count vectorizer example
from sklearn.feature_extraction.text import CountVectorizer
# start with a simple example
train_simple = ['call you tonight',
                'Call me a cab',
                'please call me... PLEASE 44!']

# learn the 'vocabulary' of the training data
vect = CountVectorizer()
train_simple_dtm = vect.fit_transform(train_simple)
pd.DataFrame(train_simple_dtm.toarray(), columns=vect.get_feature_
names())
```

	44	cab	call	me	please	tonight	you
0	0	0	1	0	0	1	1
1	0	1	1	1	0	0	0
2	1	0	1	1	2	0	0

Note that each row represents one of the three documents (sentences), each column represents one of the words present in the documents and each cell contains the number of times each word appears in each document.

We can then use the count vectorizer to transform new incoming test documents to conform with our training set (the three sentences):

```
# transform testing data into a document-term matrix (using existing
vocabulary, notice don't is missing)
test_simple = ["please don't call me"]
test_simple_dtm = vect.transform(test_simple)
test_simple_dtm.toarray()
pd.DataFrame(test_simple_dtm.toarray(), columns=vect.get_feature_
names())
```

	44	cab	call	me	please	tonight	you
0	0	0	1	1	1	0	0

Note how in our test sentence we had a new word, namely *don't*. When we vectorized it, because we hadn't seen that word previously in our training data, the vectorizer simply ignored it. This is important, and incentivizes data scientists to obtain as much data as possible for their training sets.

Now let's do this for our actual data:

```
# split into training and testing sets
from sklearn.cross_validation import train_test_split
X_train, X_test, y_train, y_test = train_test_split(df.msg, df.label,
random_state=1)


# instantiate the vectorizer
vect = CountVectorizer()
```

```
# learn vocabulary and create document-term matrix in a single step
train_dtm = vect.fit_transform(X_train)
train_dtm

<4179x7456 sparse matrix of type '<type 'numpy.int64'>'
```

With 55209 stored elements in compressed sparse row format.

Note that the format is in a sparse matrix, meaning the matrix is so large and full of zeroes, there exists a special format to deal with objects such as this. Take a look at the number of columns.

7,456 words!!

This means that in our training set, there are 7,456 unique words to look at. We can now transform our test data to conform to our vocabulary:

```
# transform testing data into a document-term matrix
test_dtm = vect.transform(X_test)
test_dtm

<1393x7456 sparse matrix of type '<type 'numpy.int64'>'
```

With 17604 stored elements in compressed sparse row format.

Note that we have the same exact number of columns because it is conforming to our test set to be exactly the same vocabulary as before. No more, no less.

Now let's build a Naïve Bayes model (similar to the linear regression process):

```
## MODEL BUILDING WITH NAIVE BAYES

# train a Naive Bayes model using train_dtm
from sklearn.naive_bayes import MultinomialNB
# import our model

nb = MultinomialNB()
# instantiate our model

nb.fit(train_dtm, y_train)
# fit it to our training set
```

Now the variable nb holds our fitted model. The training phase of the model involves computing the likelihood function, which is the conditional probability of each feature given each class:

```
# make predictions on test data using test_dtm
preds = nb.predict(test_dtm)

preds

array(['ham', 'ham', 'ham', ..., 'ham', 'spam', 'ham'],
      dtype='|S4')
```

The prediction phase of the model involves computing the posterior probability of each class given the observed features, and choosing the class with the highest probability.

We will use sklearn's built-in accuracy and confusion matrix to look at how well our Naïve Bayes models are performing:

```
# compare predictions to true labels
from sklearn import metrics
print metrics.accuracy_score(y_test, preds)
print metrics.confusion_matrix(y_test, preds)

accuracy == 0.988513998564
confusion matrix ==
[[1203    5]
 [  11  174]]
```

First off, our accuracy is great! Compared to our null accuracy which was 87%, 99% is a fantastic improvement.

Now to our confusion matrix. From before, we know that each row represents actual values while columns represent predicted values so the top left value, 1,203, represents our true negatives. But what is negative and positive? We gave the model the strings spam and ham as our classes, not positive and negative.

We can use the following:

```
nb.classes_
array(['ham', 'spam'])
```

We can then line up the indices so that the 1,203 refers to true ham predictions and 174 refers to true spam predictions.

There were also five *false spam classifications*, meaning that five messages were predicted as `spam`, but were actually `ham`, as well as 11 *false ham classifications*.

In summary, Naïve Bayes classification uses Bayes theorem in order to fit posterior probabilities of classes so that data points are correctly labeled as belonging to the proper class.

Decision trees

Decision trees are supervised models that can either preform regression or classification.

Let's take a look at some major league baseball player data from 1986-1987. Each dot represents a single player in the league:

- **Years** (x axis): Number of years played in the major leagues
- **Hits** (y axis): Number of hits the player had in the previous year
- **Salary** (color): Low salary is blue/green, high salary is red/yellow

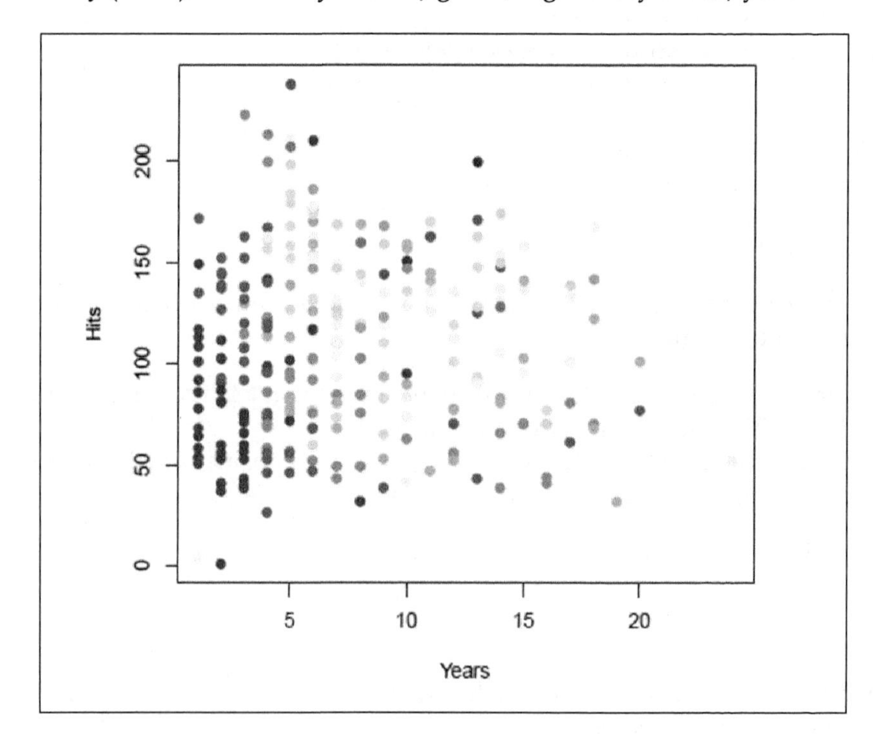

The preceding data is our training data. The idea is to build a model that predicts the salary of future players based on **Years** and **Hits**. A decision tree aims to make *splits* on our data in order to segment the data points that act similarly to each other, but differently to the others. The tree makes multiples of these splits in order to make the most accurate prediction possible. Let's see a tree built for the preceding data:

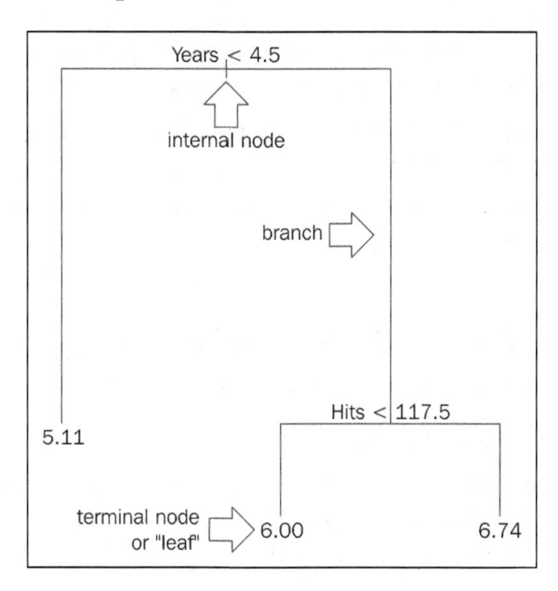

Reading from top to bottom:

- The first split is **Years < 4.5**, when a splitting rule is *true*, you follow the left branch. When a splitting rule is *false*, you follow the right branch. So for a new player, if they have been playing for less than 4.5 years, we will go down the left branch.
- For players in the left branch, the mean salary is $166,000, thus you label it with that value (salary has been divided by 1000 and log-transformed to 5.11 for ease of computation).
- For players in the right branch, there is a further split on **Hits < 117.5**, dividing players into two more salary regions: $403,000 (transformed to 6.00), and $846,000 (transformed to 6.74).

This tree doesn't just give us predictions; it also implies some more information about our data:

- It seems that the number of years in the league is the most important factor in determining salary, with a smaller number of years correlating to a lower salary
- If a player has not been playing for long (< 4.5 years), the number of hits they have is not an important factor when it comes to their salary
- For players with 5+ years under their belt, hits are an important factor for their salary determination
- Our tree only made up to two decisions before spitting out an answer (two is called our depth of the tree)

How does a computer build a regression tree?

Modern decision tree algorithms tend to use a recursive binary splitting approach:

1. The process begins at the top of the tree.
2. For every feature, it will examine every possible split, and choose the feature and split such that the resulting tree has the lowest possible mean squared error (MSE). The algorithm makes that split.
3. It will then examine the two resulting regions, and again make a single split (in one of the regions) to minimize the MSE.
4. Keep repeating step 3 until a stopping criterion is met:
    - Maximum tree depth (maximum number of splits required to arrive at a leaf)
    - Minimum number of observations in a leaf (final) node

For classification trees, the algorithm is very similar with the biggest difference being the metric we optimize over. Because MSE only exists for regression problems, we cannot use it. However instead of accuracy, classification trees optimize over either the **gini index** or **entropy**.

How does a computer fit a classification tree?

Similarly to a regression tree, a classification tree is built by optimizing over a metric (in this case, the gini index) and choosing the best split to make this optimization. More formally, at each node the tree will take the following steps:

1. Calculate the purity of the data.
2. Select a candidate split.

3. Calculate the purity of the data after the split.
4. Repeat for all variables.
5. Choose the variable with the greatest increase in purity.
6. Repeat for each split until some stop criteria is met.

Let's say that we are predicting the likelihood of death aboard a luxury cruise ship given demographic features. Suppose we start with 25 people, 10 of whom survived, and 15 of whom died:

Before Split	All
Survived	10
Died	15

We first calculate the gini index before doing anything:

$$1-\sum\left(\frac{class_i}{total}\right)^2$$

Overall classes (in this case, survived and died):

$$1-\left(\frac{survived}{total}\right)^2-\left(\frac{died}{total}\right)^2$$

$$1-\left(\frac{10}{25}\right)^2-\left(\frac{15}{25}\right)^2=0.48$$

This means that the purity of the dataset is *0.48*.

Now let's consider a potential split on gender: We first calculate the gini index for each gender:

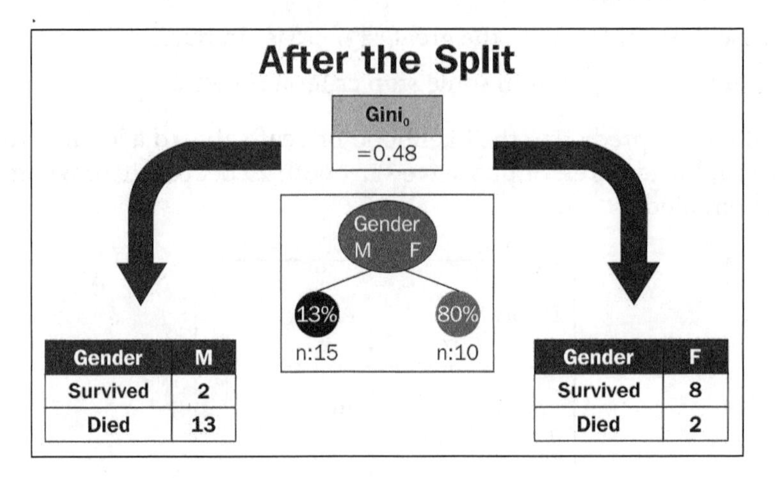

$$\text{gini(m)} = 1-\left(\frac{2}{15}\right)^2+\left(\frac{13}{15}\right)^2=.23$$

$$\text{gini(f)} = 1-\left(\frac{8}{10}\right)^2+\left(\frac{2}{10}\right)^2=.32$$

Once we have the gini index for EACH gender, we then calculate the overall gini index for the split on gender, as follows:

$$Gini(M)(M/M+F)+Gini(F)(F/M+F)=0.23(15/10+15)+0.32(10/10+15)=0.27$$

So the gini coefficient for splitting on gender is **0.27**. We then follow this procedure for three potential splits:

- Gender (male or female)
- Number of siblings on board (0 or 1+)
- Class (first and second versus third)

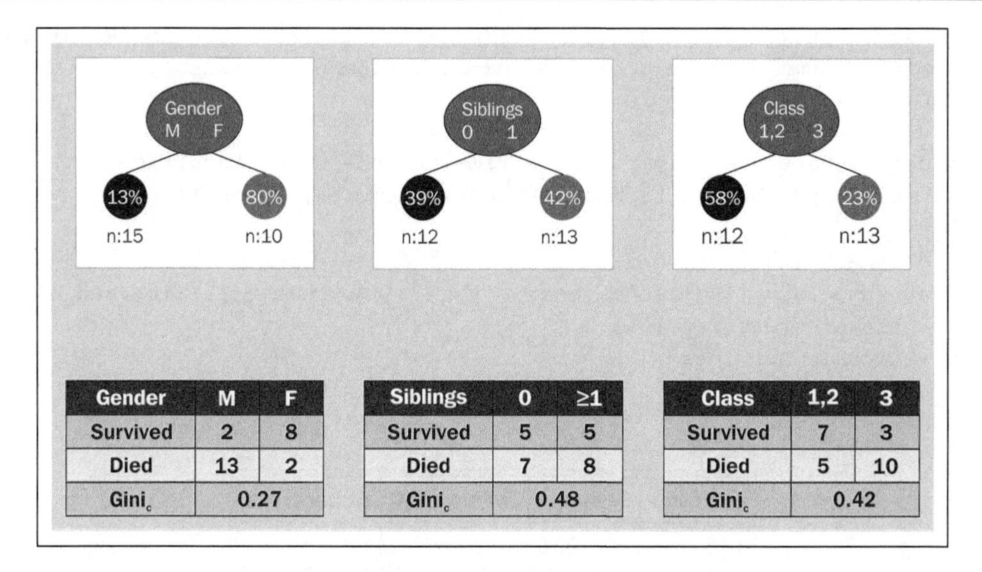

In this example, we would choose gender to split on as it is the lowest gini index!

The following table briefly summarizes the differences between classification and regression decision trees:

Regression trees	**Classification trees**
Predict a quantitative response	Predict a qualitative response
Prediction is the average value in each leaf	Prediction is the most common label in each leaf
Splits are chosen to minimize MSE	Splits are chosen to minimize gini index (usually)

Let's use scikit-learn's builtin decision tree in order to build a decision tree:

```
# read in the data
titanic = pd.read_csv('titanic.csv')

# encode female as 0 and male as 1
titanic['Sex'] = titanic.Sex.map({'female':0, 'male':1})

# fill in the missing values for age with the median age
titanic.Age.fillna(titanic.Age.median(), inplace=True)

# create a DataFrame of dummy variables for Embarked
```

```
embarked_dummies = pd.get_dummies(titanic.Embarked, prefix='Embarked')
embarked_dummies.drop(embarked_dummies.columns[0], axis=1,
inplace=True)

# concatenate the original DataFrame and the dummy DataFrame
titanic = pd.concat([titanic, embarked_dummies], axis=1)

# define X and y
feature_cols = ['Pclass', 'Sex', 'Age', 'Embarked_Q', 'Embarked_S']
X = titanic[feature_cols]
y = titanic.Survived

X.head()
```

	Pclass	Sex	Age	Embarked_Q	Embarked_S
0	3	1	22.0	0.0	1.0
1	1	0	38.0	0.0	0.0
2	3	0	26.0	0.0	1.0
3	1	0	35.0	0.0	1.0
4	3	1	35.0	0.0	1.0

Note that we are going to use class, sex, age, and dummy variables for city embarked as our features:

```
# fit a classification tree with max_depth=3 on all data
from sklearn.tree import DecisionTreeClassifier
treeclf = DecisionTreeClassifier(max_depth=3, random_state=1)
treeclf.fit(X, y)
```

`max_depth` is a limit to the depth of our tree. It means that for any data point, our tree is only able to ask up to three questions and make up to three splits. We can output our tree into a visual format and we will obtain the following:

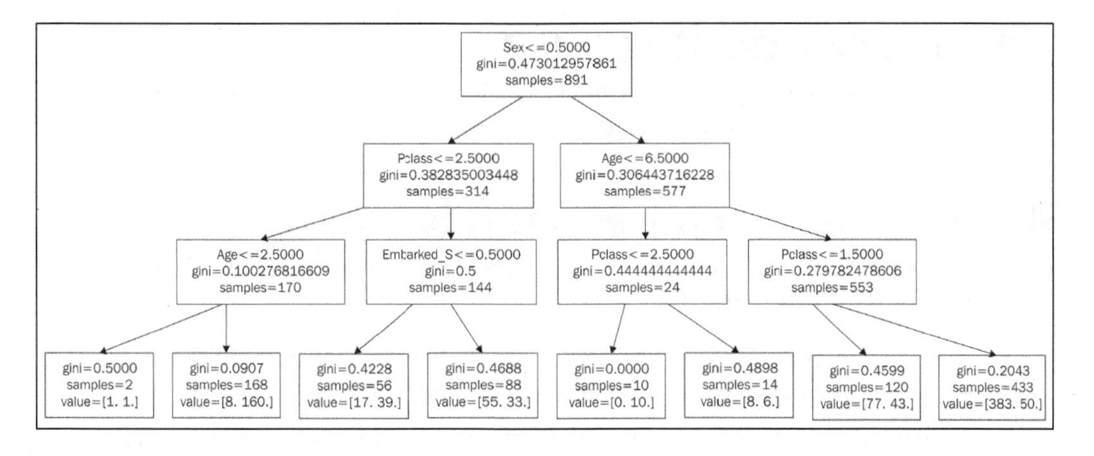

We can notice a few things:

- **Sex** is the first split, meaning that sex is the most important determining factor of whether or not a person survived the crash
- **Embarked_Q** was never used in any split

For either classification or regression trees, we can also do something very interesting with decision trees, which is that we can output a number that represents each feature's importance in the prediction of our data points:

```
# compute the feature importances
pd.DataFrame({'feature':feature_cols, 'importance':treeclf.feature_
importances_})
```

	feature	importance
0	Pclass	0.242664
1	Sex	0.655584
2	Age	0.064494
3	Embarked_Q	0.000000
4	Embarked_S	0.037258

The importance scores are an average gini index difference for each variable, with higher values corresponding to higher *importance* to the prediction. We can use this information to select fewer features in the future. For example, both of the embarked variables are very low in comparison to the rest of the features, so we may be able to say that they are not important in our prediction of life or death.

Unsupervised learning

It's time to see some examples of unsupervised learning, given that we spend a majority of this book on **supervised learning models**.

When to use unsupervised learning

There are many times when unsupervised learning can be appropriate. Some very common examples include the following:

- When there is no clear *response* variable. There is nothing that we are explicitly trying to predict or correlate to other variables.
- To extract structure from data where no apparent structure/patterns exist (can be a supervised learning problem).
- When an unsupervised concept called feature extraction is used. Feature extraction is the process of creating new features from existing ones. These new features can be even stronger than the original features.

The first tends to be the most common reason that data scientists choose to use unsupervised learning. This case arises frequently when we are working with data and we are not explicitly trying to predict any of the columns and we merely wish to find patterns of similar (and dissimilar) groups of points. The second option comes into play even if we are explicitly attempting to use a supervised model to predict a response variable. Sometimes simple EDA might not produce any clear patterns in the data in the few dimensions that humans can imagine where as a machine might pick up on data points behaving similarly to each other in greater dimensions.

The third common reason to use unsupervised learning is to extract new features from features that already exist. This process (*lovingly called feature extraction*) might produce features that can be used in a future supervised model or that can be used for presentation purposes (marketing or otherwise).

K-means clustering

K-means clustering is our first example of an unsupervised machine learning model. Remember this means that we are not making predictions; we are trying instead to extract structure from seemingly unstructured data.

Clustering is a family of unsupervised machine learning models that attempt to group data points into **clusters** with **centroids**.

Definition:

Cluster: A group of data points that behave similarly.

Definition:

Centroid: The *center* of a cluster. Can be thought of as an *average* point in the cluster.

The preceding definition can be quite vague, but it becomes specific when narrowed down to specific domains. For example, online shoppers who behave similarly might shop for similar things or at similar shops, whereas similar software companies might make comparable software at comparable prices.

Here is a visualization of clusters of points:

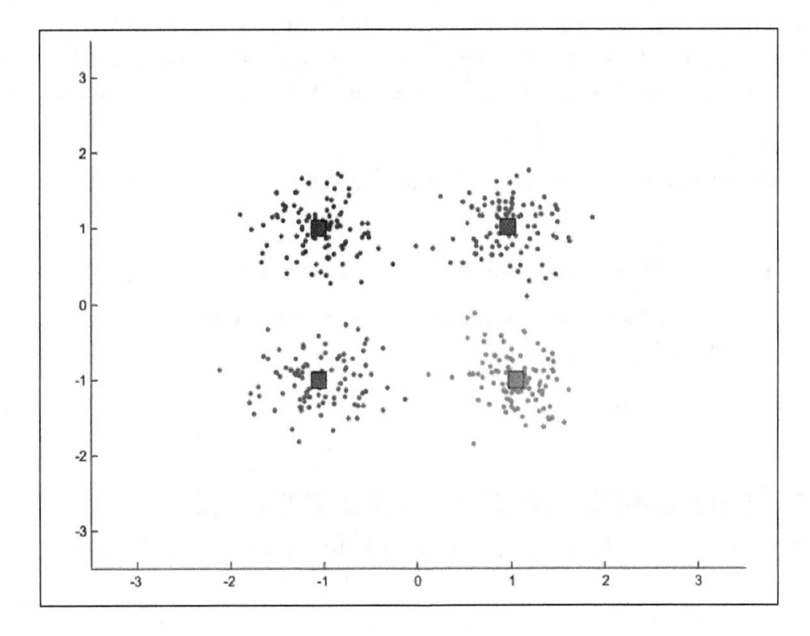

In the preceding figure, our human brains can very easily see the difference between the four clusters. Namely that the red cluster is on the bottom left of the graph while the green cluster lives in the bottom right portion of the graph. This means that the red data points are similar to each other, but *not* similar to data points in the other clusters.

We can also see the centroids of each cluster as the square in each color. Note that the centroid is *not* an actual data point, but is merely an abstraction of a cluster and represents the center of the cluster.

The concept of *similarity* is central to the definition of a cluster, and therefore to cluster analysis. In general, greater similarity between points leads to better clustering. In most cases, we turn data into points in *n*-dimensional space and use the distance between these points as a form of similarity. The centroid of the cluster then is usually the average of each dimension (column) for each data point in each cluster. So for example, the centroid of the red cluster is the result of taking the average value of each column of each red data point.

The purpose of cluster analysis is to enhance our understanding of a dataset by dividing the data into groups. Clustering provides a layer of abstraction from individual data points. The goal is to extract and enhance the natural structure of the data. There are many kinds of classification procedures. For our class, we will be focusing on K-means clustering, which is one of the most popular clustering algorithms.

K-means is an iterative method that partitions a data set into `k` clusters. It works in four steps:

1. Choose `k` initial centroids (note that `k` is an input).
2. For each point assign the point to the nearest centroid.
3. Recalculate the centroid positions.
4. Repeat steps 2-3 until stopping criteria is met.

Illustrative example – data points

Imagine that we have the following data points in a two-dimensional space:

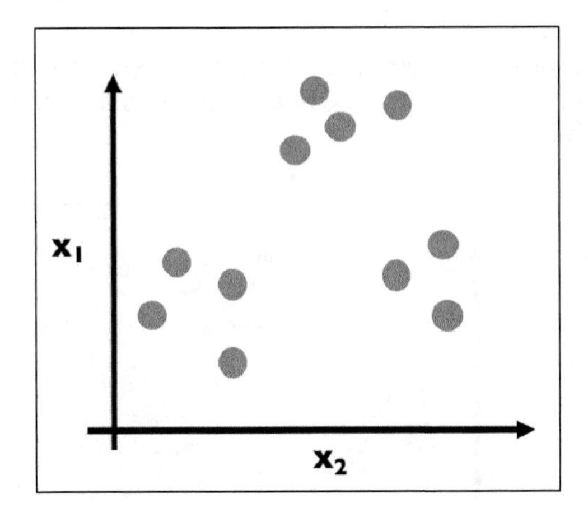

Each dot is colored grey so as to assume no prior grouping before applying the K-means algorithm. The goal here is to eventually color in each dot and create groupings (clusters).

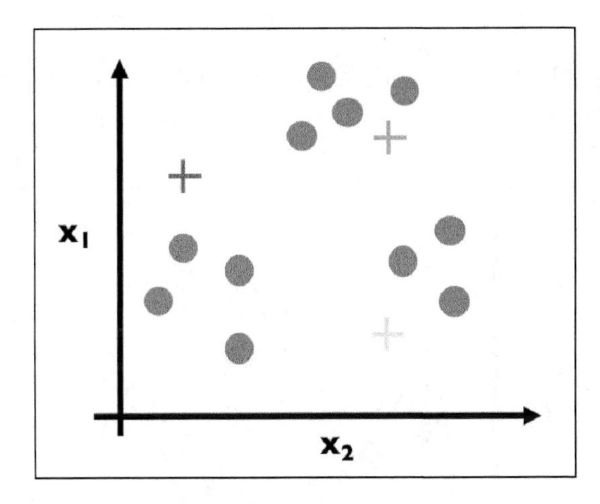

Here, step 1 has been applied. We have (randomly) chosen three centroids (red, blue, and yellow).

Most K-means algorithms place random initial centroids, but there exist other pre-compute methods to place initial centroids. For now, random is fine.

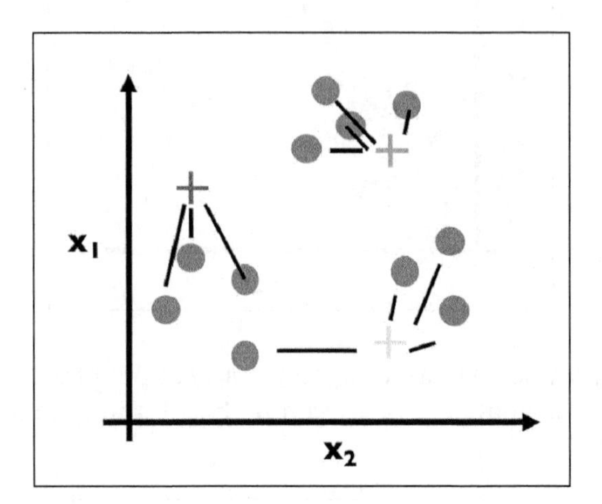

The first part of step 2 has been applied. For each data point, we found the most similar centroid (closest).

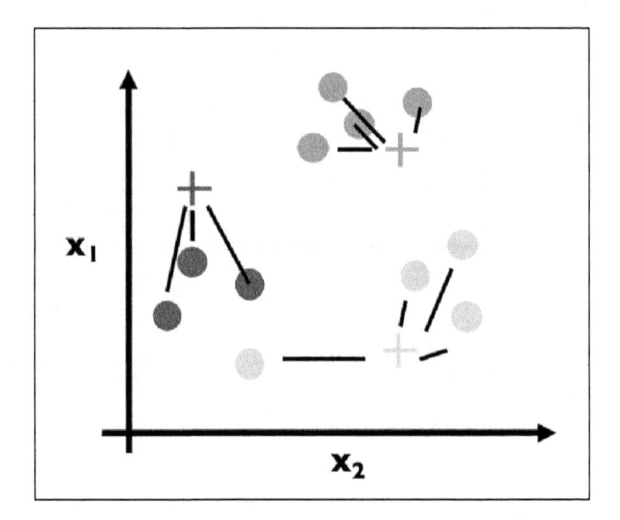

The second part of step 2 has been applied here. We have colored in each data point in accordance with its most similar centroid.

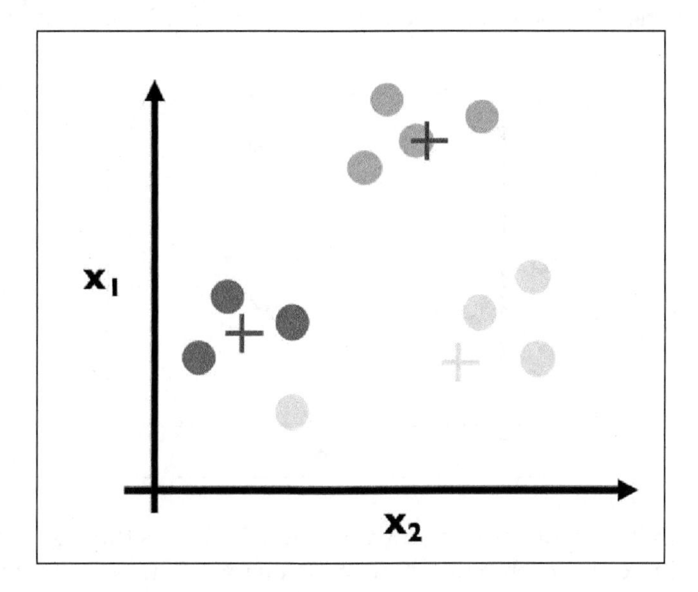

This is step 3 and the crux of K-Means. Note that we have physically moved the centroids to be the actual center of each cluster. We have, for each color, computed the average point and made that point the new centroid. For example, suppose the three red data points had the coordinates (`1, 3`), (`2, 5`), and (`3, 4`). The center (*red cross*) would be calculated as follows:

```
# centroid calculation
import numpy as np
red_point1 = np.array([1, 3])
red_point2 = np.array([2, 5])
red_point3 = np.array([3, 4])

red_center = (red_point1 + red_point2 + red_point3) / 3.

red_center
# array([ 2.,  4.])
```

That is, the point (`2, 4`) would be the coordinates of the preceding red cross.

None of the actual data points will ever move. They cannot. The only entities that move are the centroids, which are NOT actual data points.

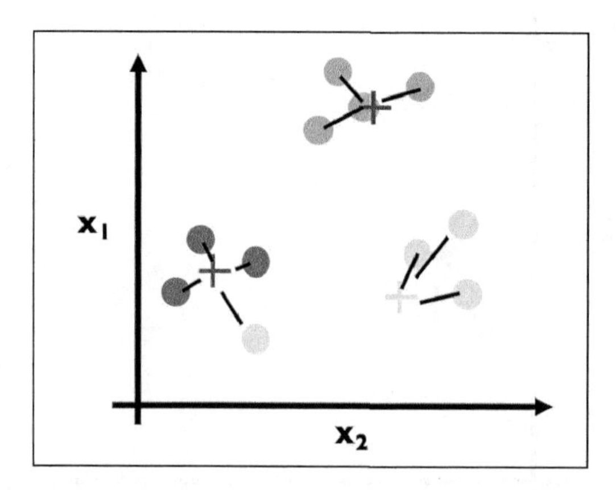

We continue with our algorithm by repeating step 2. Here is the first part where we find the closest center for each point. Note a big change: The point that is circled in the following figure used to be a yellow point, but has changed to be a red cluster point because the yellow cluster moved closer to its yellow constituents.

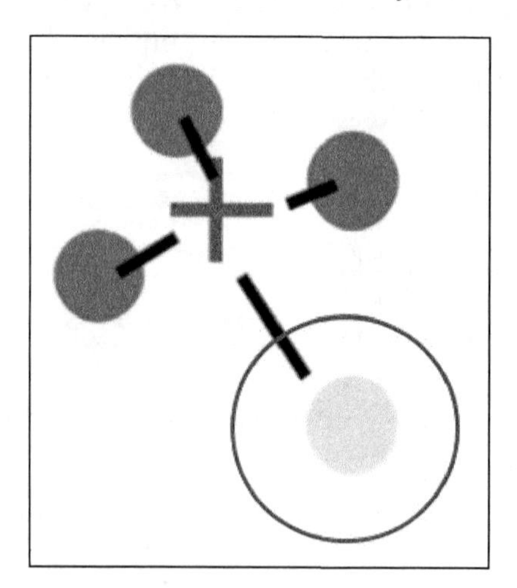

It might help to think of points as being planets in space with gravitational pull. Each centroid is pulled by the planets' gravities.

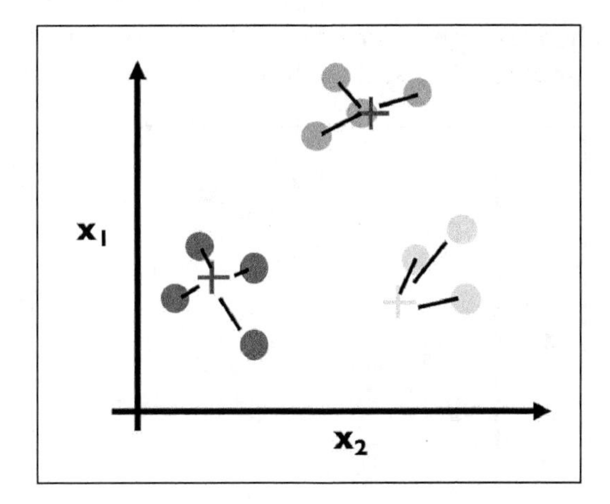

Here is the second part of step 2 again. We have assigned each point to the color of the closest cluster.

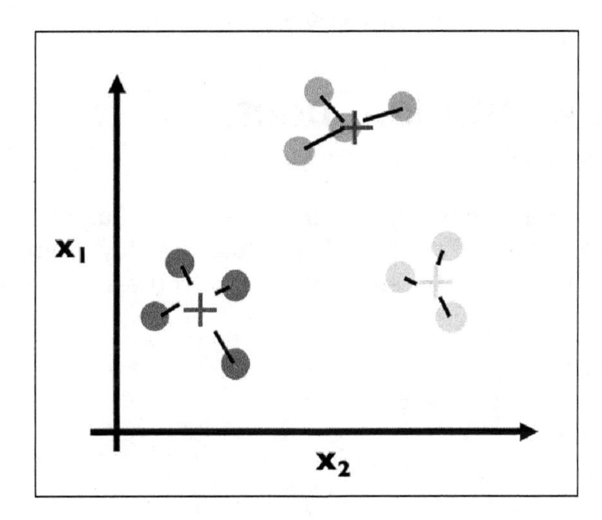

Here, we recalculate once more the centroids for each cluster (step 3). Note that the blue center did not move at all, while the yellow and red centers both moved.

Because we have reached a stopping criterion (clusters do not move if we repeat step 2 and 3), we finalize our algorithm and we have our three clusters!

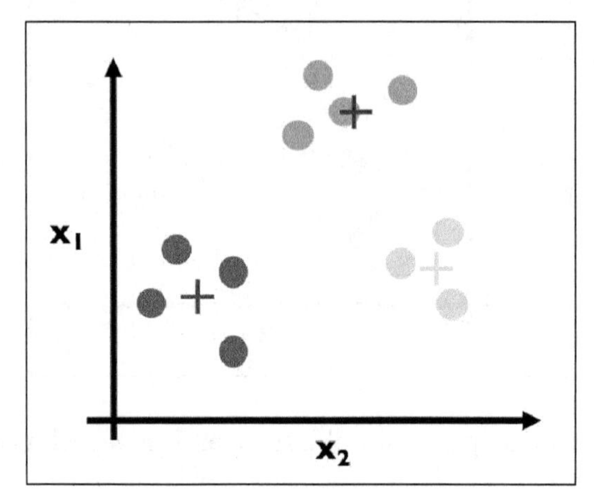

Final results of K-means algorithm

Illustrative example – beer!

Enough data science, beer!

Ok ok, settle down. It's a long book, let's grab a beer. On that note, did you know there are many types of beer? I wonder if we could possibly group beers into different categories based on different quantitative features... Let's try!

```
# import the beer dataset
url = '../data/beer.txt'
beer = pd.read_csv(url, sep=' ')
print beer.shape
(20, 5)

beer.head()
```

	name	calories	sodium	alcohol	cost
0	Budweiser	144	15	4.7	0.43
1	Schlitz	151	19	4.9	0.43
2	Lowenbrau	157	15	0.9	0.48
3	Kronenbourg	170	7	5.2	0.73
4	Heineken	152	11	5.0	0.77

Here we have 20 beers with five columns: name, calories, sodium, alcohol, and cost. In clustering (like almost all machine learning models), we like quantitative features, so we will ignore the name of the beer in our clustering:

```
# define X
X = beer.drop('name', axis=1)
```

Now we will perform K-Means using scikit-learn:

```
# K-means with 3 clusters
from sklearn.cluster import KMeans
km = KMeans(n_clusters=3, random_state=1)
km.fit(X)
```

`n_clusters` is our `k`. It is our inputted number of clusters. `random_state` as always produces reproducible results for educational purposes. Using three clusters for now is random.

Our K-means algorithm has run the algorithm on our data points and come up with three clusters:

```
# save the cluster labels and sort by cluster
beer['cluster'] = km.labels_
```

We can take a look at the center of each cluster by using a `groupby` and `mean` statement:

```
# calculate the mean of each feature for each cluster
beer.groupby('cluster').mean()
```

	calories	sodium	alcohol	cost
cluster				
0	150.00	17.0	4.521429	0.520714
1	102.75	10.0	4.075000	0.440000
2	70.00	10.5	2.600000	0.420000

On human inspection, we can see that cluster 0 has, on average, a higher calorie content, sodium content, and alcohol content, and costs more. These might be considered heavier beers. Cluster 2 has on average a very low alcohol content and very few calories. These are probably light beers. Cluster 1 is somewhere in the middle.

Let's use Python to make a graph to see this in more detail:

```
import matplotlib.pyplot as plt
%matplotlib inline


# save the DataFrame of cluster centers
centers = beer.groupby('cluster').mean()
# create a "colors" array for plotting
colors = np.array(['red', 'green', 'blue', 'yellow'])
# scatter plot of calories versus alcohol, colored by cluster (0=red,
1=green, 2=blue)
plt.scatter(beer.calories, beer.alcohol, c=colors[list(beer.cluster)],
s=50)

# cluster centers, marked by "+"
plt.scatter(centers.calories, centers.alcohol, linewidths=3,
marker='+', s=300, c='black')

# add labels
plt.xlabel('calories')
plt.ylabel('alcohol')
```

A big part of unsupervised learning is human inspection. Clustering has no context of the problem domain and can only tell us the clusters it found, it cannot tell us what the clusters mean.

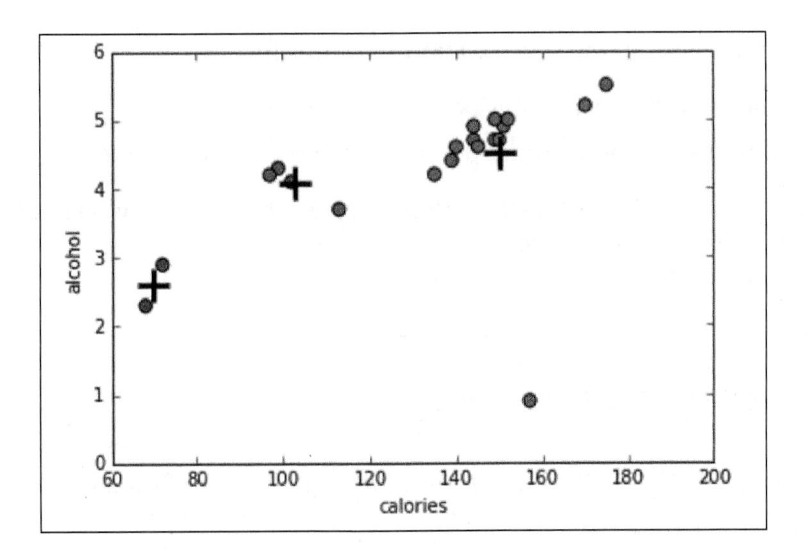

Choosing an optimal number for K and cluster validation

A big part of K-means clustering is knowing the optimal number of clusters. If we knew this number ahead of time, then that might defeat the purpose of even using unsupervised learning. So we need a way to evaluate the output of our cluster analysis.

The problem here is that because we are not performing any kind of prediction, we cannot gauge how *right* the algorithm is at predictions. Metrics such as accuracy and RMSE go right out of the window.

The Silhouette Coefficient

The **Silhouette Coefficient** is a common metric for evaluating clustering performance in situations when the *true* cluster assignments are not known.

A Silhouette Coefficient is calculated for each observation as follows:

$$SC = \frac{b-a}{\max(a,b)}$$

Let's look a little closer at the specific features of this formula:

- *a*: Mean distance to all other points in its cluster
- *b*: Mean distance to all other points in the next nearest cluster

It ranges from -1 (worst) to 1 (best). A **global score** is calculated by taking the mean score for all observations. In general, a silhouette coefficient of 1 is preferred, while a score of -1 is not preferable:

```
# calculate Silhouette Coefficient for K=3
from sklearn import metrics
metrics.silhouette_score(X, km.labels_)
0.4578
```

Let's try calculating the coefficient for multiple values of K to find the best value:

```
# calculate SC for K=2 through K=19
k_range = range(2, 20)
scores = []
for k in k_range:
    km = KMeans(n_clusters=k, random_state=1)
    km.fit(X_scaled)
    scores.append(metrics.silhouette_score(X, km.labels_))

# plot the results
plt.plot(k_range, scores)
plt.xlabel('Number of clusters')
plt.ylabel('Silhouette Coefficient')
plt.grid(True)
```

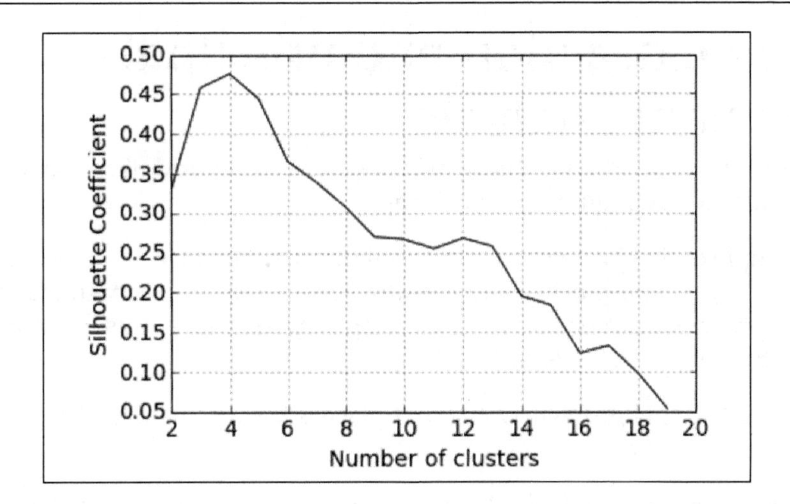

So it looks like our optimal number of beer clusters is 4! This means that our k-means algorithm has determined that there seems to be four distinct types of beer.

K-means is a popular algorithm because of its computational efficiency and simple and intuitive nature. K-means, however, is highly scale dependent, and is not suitable for data with widely varying shapes and densities. There are ways to combat this issue by scaling data using scikit-learn's standard scalar:

```
# center and scale the data
from sklearn.preprocessing import StandardScaler
scaler = StandardScaler()
X_scaled = scaler.fit_transform(X)

# K-means with 3 clusters on scaled data
km = KMeans(n_clusters=3, random_state=1)
km.fit(X_scaled)
```

Easy!

Now let's take a look at the third reason to use unsupervised methods that falls under the third option in our reasons to use unsupervised methods, feature extraction.

Feature extraction and principal component analysis

Sometimes we have an overwhelming number of columns and likely not enough rows to handle the great quantity of columns.

A great example of this is when we were looking at the `send cash now` example in our Naïve Bayes example. We had literally 0 instances of texts with that exact phrase, so instead we turned to a *naïve* assumption that allowed us to extrapolate a probability for both of our categories.

The reason we had this problem in the first place is because of something called the curse of dimensionality.

The curse of dimensionality basically says that as we introduce and consider new feature columns, we need almost exponentially more rows (data points) in order to fill in the empty spaces that we create.

Consider an example where we attempt to use a learning model that utilizes the distance between points on a corpus of text that has 4,086 pieces of text, and that the whole thing has been `Countvectorized`. Let's assume that these texts between them have 18,884 words:

```
X.shape
(4086, 18884)
```

Now let's do an experiment. I will first consider a single word as the only dimension of our text. Then I will count how many of pieces of text are within 1 *unit* of each other. For example, if two sentences both contain that word, they would be 0 units away and similarly if neither of them contain the word, they would be 0 units away from one another:

```
d = 1
# Let's look for points within 1 unit of one another

X_first_word = X[:,:1]
# Only looking at the first column, but ALL of the rows

from sklearn.neighbors import NearestNeighbors
# this module will calculate for us distances between each point

neigh = NearestNeighbors(n_neighbors=4086)
neigh.fit(X_first_word)
# tell the module to calculate each distance between each point
```

Note that we have 16,695,396 (4086*4086) distances to scan over

```
A = neigh.kneighbors_graph(X_first_word, mode='distance').todense()
# This matrix holds all distances (over 16 million of them)

num_points_within_d = (A < d).sum()
# Count the number of pairs of points within 1 unit of distance

num_points_within_d
16258504
```

So 16.2 million pairs of texts are within a single unit of distance. Now let's try again with the first two words:

```
X_first_two_words = X[:,:2]
neigh = NearestNeighbors(n_neighbors=4086)
neigh.fit(X_first_two_words)
A = neigh.kneighbors_graph(X_first_two_words, mode='distance').
todense()
num_points_within_d = (A < d).sum()

num_points_within_d
16161970
```

Great! By adding this new column, we lost about 100,000 pairs of points that were within a single unit of distance. This is because we are adding space in between them for every dimension that we add. Let's take this test a step further and calculate this number for the first 100 words and then plot the results:

```
d = 1
# Scan for points within one unit

num_columns = range(1, 100)
# Looking at the first 100 columns
points = []
# We will be collecting the number of points within 1 unit for a graph

neigh = NearestNeighbors(n_neighbors=X.shape[0])
for subset in num_columns:
    X_subset = X[:,:subset]
  # look at the first column, then first two columns, then first three
columns, etc
    neigh.fit(X_subset)
```

```
        A = neigh.kneighbors_graph(X_subset, mode='distance').todense()
        num_points_within_d = (A < d).sum()
    # calculate the number of points within 1 unit
        points.append(num_points_within_d)
```

Now let's plot the number of points within 1 unit versus the number of dimensions we looked at:

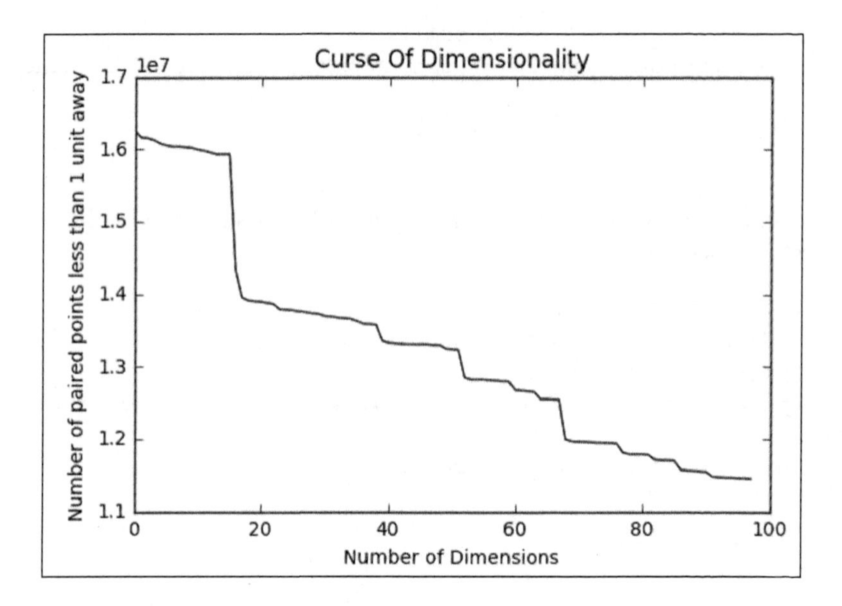

We can see clearly that the number of points within a single unit of one another goes down dramatically as we introduce more and more columns. And this is only the first 100 columns! Let's see how many points are within a single unit by the time we consider all 18,000+ words:

```
neigh = NearestNeighbors(n_neighbors=4086)
neigh.fit(X)
A = neigh.kneighbors_graph(X, mode='distance').todense()
num_points_within_d = (A < d).sum()


num_points_within_d
4090
```

By the end, only 4,000 sentences are within a unit of one another. All of this space that we add in by considering new columns makes it harder for the finite amount of points we have to stay happily within range of each other. We would have to add in more points in order to fill in this gap. And that, my friends, is why we should consider using dimension reduction.

The curse of dimensionality is solved by either adding more data points (which is not always possible), or implementing dimension reduction. Dimension reduction is simply the act of reducing the number of columns in our data set and not the number of rows. There are two ways of implementing dimension reduction:

- **Feature selection**: This is the act of subsetting our column features and only using the best features
- **Feature extraction**: This is the act of mathematically transforming our feature set into a new extracted coordinate system

We are familiar with feature selection as the process of saying *the Emabrked_Q is not helping my decision tree; let's get rid of it and see how it performs*. It is literally when we (or the machine) make the decision to ignore certain columns.

Feature extraction is a bit trickier...

In feature extraction, we are using usually fairly complicated mathematical formulas in order to obtain new *super-columns* that are usually better than any single original column.

Our primary model for doing so is called **Principal Component Analysis** (**PCA**). PCA will extract a set number of super-columns in order to represent our original data with much fewer columns. Let's take a concrete example. Previously I mentioned some text with 4,086 rows and over 18,000 columns. That dataset is actually a set of Yelp online reviews:

```
url = '../data/yelp.csv'
yelp = pd.read_csv(url, encoding='unicode-escape')

# create a new DataFrame that only contains the 5-star and 1-star
reviews
yelp_best_worst = yelp[(yelp.stars==5) | (yelp.stars==1)]

# define X and y
X = yelp_best_worst.text
y = yelp_best_worst.stars == 5
```

Our goal is to predict whether or not a person gave a 5 or 1 star review based on the words they used in the review. Let's set a base line with logistic regression and see how well we can predict this binary category:

```
from sklearn.linear_model import LogisticRegression
lr = LogisticRegression()

X_train, X_test, y_train, y_test = train_test_split(X, y, random_
state=100)
```

```
# Make our training and testing sets

vect = CountVectorizer(stop_words='english')
# Count the number of words but remove stop words like a, an, the,
you, etc

X_train_dtm = vect.fit_transform(X_train)
X_test_dtm = vect.transform(X_test)
# transform our text into document term matrices

lr.fit(X_train_dtm, y_train)
# fit to our training set

lr.score(X_test_dtm, y_test)
# score on our testing set
0.91193737
```

So by utilizing all of the words in our corpus, our model seems to have over a 91% accuracy. Not bad!

Let's try only using the top 100 used words:

```
vect = CountVectorizer(stop_words='english', max_features=100)
# Only use the 100 most used words

X_train_dtm = vect.fit_transform(X_train)
X_test_dtm = vect.transform(X_test)
print X_test_dtm.shape  # (1022, 100)


lr.fit(X_train_dtm, y_train)

lr.score(X_test_dtm, y_test)
0.8816
```

Note how our training and testing matrices have 100 columns. This is because I told our vectorizer to only look at the top 100 words. See also that our performance took a hit and is now down to 88% accuracy. This makes sense because we are ignoring over 4,700 words in our corpus.

Now let's take a different approach. Let's import a PCA module and tell it to make us 100 NEW super-columns and see how that performs:

```
from sklearn import decomposition
# We will be creating 100 super columns
```

```
vect = CountVectorizer(stop_words='english')
# Don't ignore any words
pca  = decomposition.PCA(n_components=100)
# instantate a pca object

X_train_dtm = vect.fit_transform(X_train).todense()
# A dense matrix is required to pass into PCA, does not affect the
overall message
X_train_dtm = pca.fit_transform(X_train_dtm)

X_test_dtm = vect.transform(X_test).todense()
X_test_dtm = pca.transform(X_test_dtm)
print X_test_dtm.shape  # (1022, 100)


lr.fit(X_train_dtm, y_train)

lr.score(X_test_dtm, y_test)
.89628
```

Not only do our matrices still have 100 columns, but these columns are no longer words in our corpus. They are complex transformations of columns and are 100 new columns. Also note that using 100 of these new columns gives us a better predictive performance than using the 100 top words!

Feature extraction is a great way to use mathematical formulas to extract brand new columns that generally perform better than just selecting the best ones beforehand.

But how do we visualize these new super columns? Well I can think of no better way than to look at an example using image analysis. Specifically, let's make a facial recognition software. OK? OK. Let's begin by importing some faces given to us by scikit-learn:

```
from sklearn.datasets import fetch_lfw_people

lfw_people = fetch_lfw_people(min_faces_per_person=70, resize=0.4)

# introspect the images arrays to find the shapes (for plotting)
n_samples, h, w = lfw_people.images.shape

# for machine learning we use the 2 data directly (as relative pixel
# positions info is ignored by this model)
X = lfw_people.data
y = lfw_people.target
```

```
n_features = X.shape[1]

X.shape
(1288, 1850)
```

We have gathered 1,288 images of people's faces and each one has 1,850 features (pixels) that identify that person. For example:

```
plt.imshow(X[0].reshape((h, w)), cmap=plt.cm.gray)
lfw_people.target_names[y[0]]
'Hugo Chavez'
```

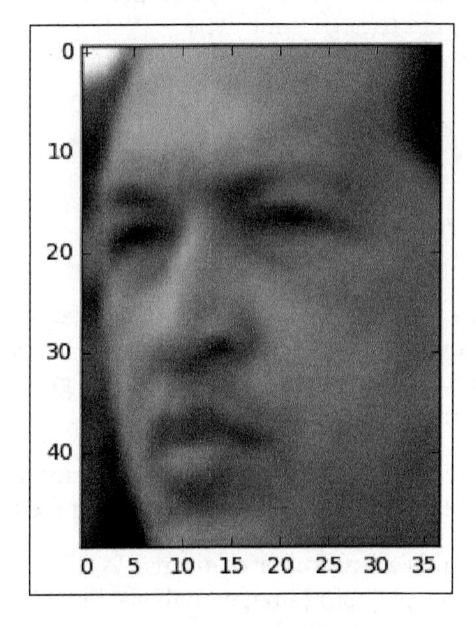


```
plt.imshow(X[100].reshape((h, w)), cmap=plt.cm.gray)
lfw_people.target_names[y[100]]
'George W Bush'
```

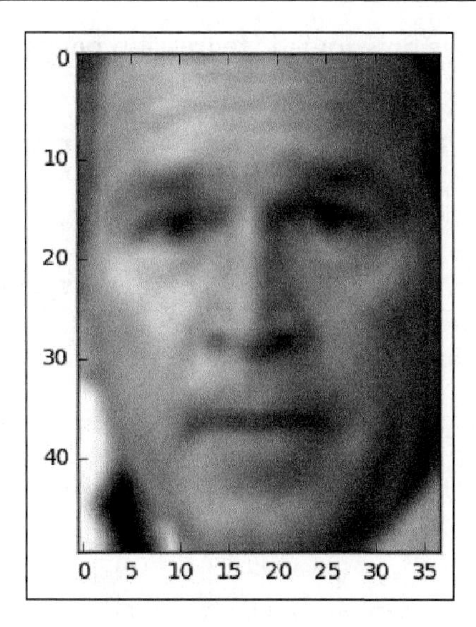

Great. To get a glimpse at the type of data set we are looking at, let's look at a few overall metrics:

```
# the label to predict is the id of the person
target_names = lfw_people.target_names
n_classes = target_names.shape[0]

print("Total dataset size:")
print("n_samples: %d" % n_samples)
print("n_features: %d" % n_features)
print("n_classes: %d" % n_classes)

Total dataset size:
n_samples: 1288
n_features: 1850
n_classes: 7
```

So we have 1,288 images, 1,850 features, and 7 classes (people) to choose from. Our goal is to make a classifier that will assign the person's face a name based on the 1,850 pixels given to us.

Let's take a base line and see how logistic regression performs on our data without doing anything:

```
from sklearn.linear_model import LogisticRegression
from sklearn.metrics import accuracy_score
from time import time  # for timing our work

X_train, X_test, y_train, y_test = train_test_split(
    X, y, test_size=0.25, random_state=1)
# get our training and test set

t0 = time() # get the time now
logreg = LogisticRegression()

logreg.fit(X_train, y_train)

# Predicting people's names on the test set
y_pred = logreg.predict(X_test)


print accuracy_score(y_pred, y_test), "Accuracy"
print (time() - t0), "seconds"

0.810559006211 Accuracy
6.31762504578 seconds
```

So within 6.3 seconds, we were able to get an 81% on our test set. Not too bad...

Now let's try this with our *super faces*:

```
# split into a training and testing set
from sklearn.cross_validation import train_test_split

# Compute a PCA (eigenfaces) on the face dataset (treated as unlabeled
# dataset): unsupervised feature extraction / dimensionality reduction
n_components = 75

# Extracting the top %d eigenfaces from %d faces
      % (n_components, X_train.shape[0]))
pca = decomposition.PCA(n_components=n_components, whiten=True).fit(X_
train)
# This whiten parameter speeds up the computation of our extracted
columns
```

```
# Projecting the input data on the eigenfaces orthonormal basis
X_train_pca = pca.transform(X_train)
X_test_pca = pca.transform(X_test)
```

The preceding code is collecting 75 extracted columns from our 1,850 unprocessed columns. These are our `super faces`. Now let's plug in our newly extracted columns into our logistic regression and compare:

```
t0 = time()

# Predicting people's names on the test set WITH PCA
logreg.fit(X_train_pca, y_train)
y_pred = logreg.predict(X_test_pca)

print accuracy_score(y_pred, y_test), "Accuracy"
print (time() - t0), "seconds"

0.82298136646 Accuracy
0.194181919098 seconds
```

Wow! Not only was this entire calculation about 30 times faster than the unprocessed images, the predictive performance got better! This shows us that PCA and feature extraction in general can help us all around when performing machine learning on complex data sets with many columns. By searching for these patterns in the dataset and extracting new feature columns, we can speed up and enhance our learning algorithms.

Let's look at one more interesting thing... I mentioned before that one of the purposes of this example was to examine and visualize our *eigenfaces*, as they are called. Our super columns. I will not disappoint. Let's write some code that will show us our super-columns as they would look like to us humans:

```
def plot_gallery(images, titles, n_row=3, n_col=4):
    """Helper function to plot a gallery of portraits"""
    plt.figure(figsize=(1.8 * n_col, 2.4 * n_row))
    plt.subplots_adjust(bottom=0, left=.01, right=.99, top=.90,
hspace=.35)
    for i in range(n_row * n_col):
        plt.subplot(n_row, n_col, i + 1)
        plt.imshow(images[i], cmap=plt.cm.gray)
        plt.title(titles[i], size=12)

# plot the gallery of the most significative eigenfaces
eigenfaces = pca.components_.reshape((n_components, h, w))
eigenface_titles = ["eigenface %d" % i for i in range(eigenfaces.
shape[0])]
```

```
plot_gallery(eigenfaces, eigenface_titles)

plt.show()
```

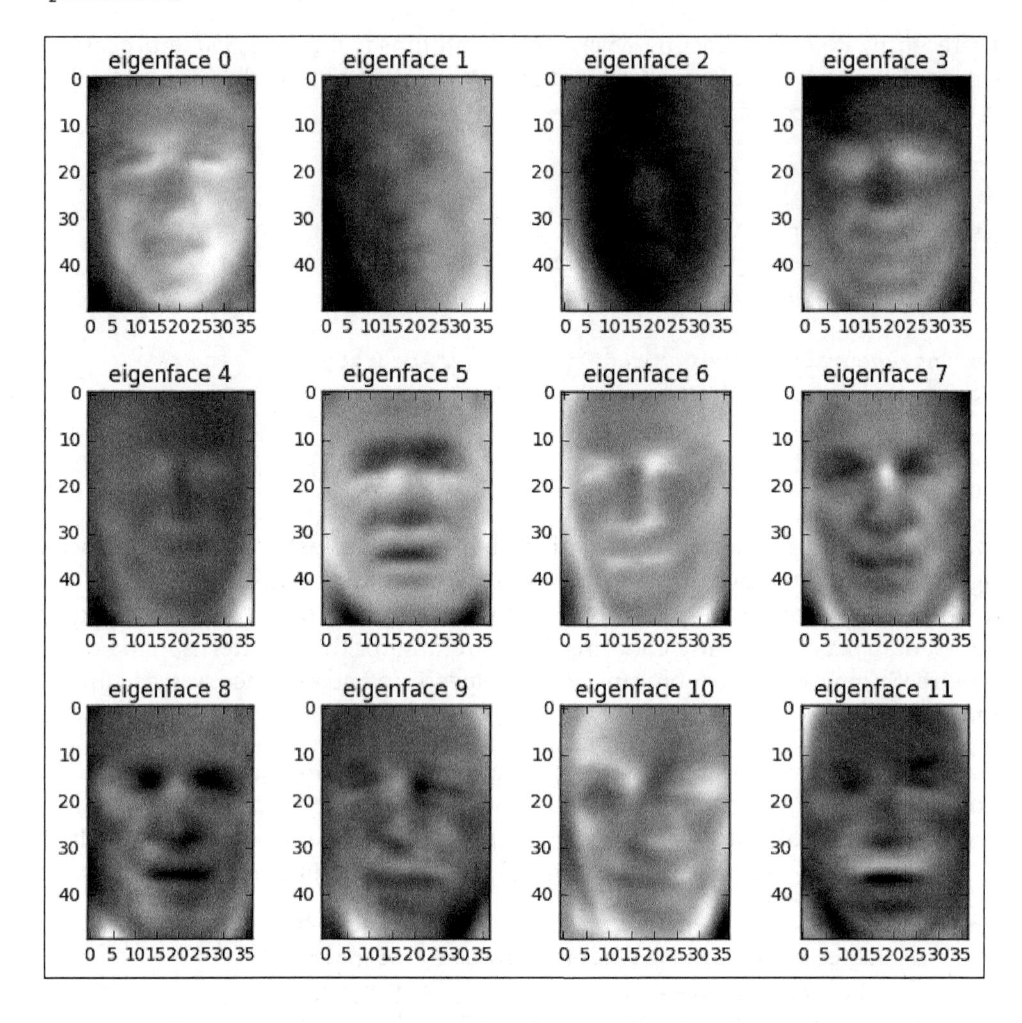

Wow. A haunting and yet beautiful representation of what the data believes to be the most importance features of a face. As we move from the top left (first super column) to the bottom, it is actually somewhat easy to see what the image is trying to tell us. The first super column looks like a very general face structure with eyes and nose and a mouth. It is almost saying "I represent the basic qualities of a face that all faces must have". Our second super column directly to its right seems to be telling us about shadows in the image. The next one might be telling us that skin tone plays a role in detecting who this is, which might be why the third face is much darker than the first two.

Using feature extraction unsupervised learning methods such as PCA can give us a very deep look into our data and reveal to us what the data believes to be the most important features, not just what we believe them to be. Feature extraction is a great preprocessing tool that can speed up our future learning methods, make them more powerful, and give us more insight into how the data believes it should be viewed. To sum up this section, we will list the pros and cons.

Pros of using feature extraction:

- Our models become much faster
- Our predictive performance can become better
- Can give us insight into the extracted features (eigenfaces)

Cons of using feature extraction:

- We lose interpretability of our features as they are new mathematically derived columns, not our old ones
- We can lose predictive performance because we are losing information as we extract fewer columns

Summary

Between decision trees, Naïve Bayes classification, feature extraction, and K-means clustering, we have seen that machine learning goes way beyond the simplicity of linear and logistic regression and can solve many types of complicated problems.

We also saw examples of both supervised and unsupervised learning and in doing so became familiar with many types of data science related problems.

In the next chapter, we will be looking at even more complicated learning algorithms including artificial neural networks, and ensembling techniques. We will also see and understand more complicated concepts in data science, including the bias-variance tradeoff, as well as the concept of overfitting.

12
Beyond the Essentials

In this chapter, we will be discussing some of the more complicated parts of data science that can put some people off. The reason for this is that data science is not all fun and machine learning. Sometimes, we have to discuss and consider theoretical and mathematical paradigms and evaluate our procedures.

This chapter will explore many of these procedures step by step so that we completely and totally understand the topics. We will be discussing topics such as the following:

- Cross-validation
- The bias variance tradeoff
- Overfitting and underfitting
- Ensembling techniques
- Random forests
- Neural networks

These are only some of the topics to be covered. At no point do I want you to be confused. I will attempt to explain each procedure/algorithm with utmost care and with many examples and visuals.

The bias variance tradeoff

We have discussed the concept of bias and variance briefly in the previous chapters. When we are discussing these two concepts, we are generally speaking of supervised learning algorithms. We are specifically talking about deriving errors from our predictive models due to bias and variance.

Error due to bias

When speaking of errors due to Bias, we are speaking of the difference between the expected prediction of our model and the actual (correct) value, which we are trying to predict. Bias, in effect, measures how far, in general, our model's predictions are from the correct value.

Think about bias as simply being the difference between a predicted value and the actual value. For example, consider that our model, represented as *F(x)*, predicts the value of *29* as follows:

$$F(29) = 88$$

Here, the value of *29* should have been predicted at *79*, then:

$$Bias(29) = 88 - 79 = 9$$

If a machine learning model tends to be very accurate in its prediction (regression or classification), then it is considered a low Bias model, whereas if the model is more often than not wrong, it is considered to be a high bias model.

Bias is a measure to judge models on the basis of *accuracy* or just how correct the model is on an average.

Error due to variance

An error due to variance is dependent on the variability of a model prediction for a given data point. Imagine that you repeat the machine learning model building process over and over. The variance is measured by looking at how much the predictions for a fixed point vary between different end results.

To imagine variance in your head, think about a population of data points. If you were to take randomized samples over and over, how drastically would your machine learning model change or fit differently each time. If the model does not change much between samples, the model would be considered a low variance model. If your model changes drastically between samples, then that model would be considered a *high variance* model.

Variance is a great measure to judge our model on the basis of generalizability. If our model has a low variance, we can expect it to behave in a certain way when set into the wild and predict values without human supervision.

Our goal is to optimize both bias and variance. Ideally, we are looking for the lowest possible variance and bias.

I find that this can be best explained using an example.

Example - comparing body and brain weight of mammals

Imagine that we are considering a relationship between the brain weight of mammals and their corresponding body weights. A hypothesis might read that there is a positive correlation between the two (as one goes up, so does the other). But how strong is this relationship? Is it even linear? Perhaps, as the brain weight increases, there is a logarithmic or quadratic increase in body weight.

Let's use Python to explore, as shown:

```
# # Exploring the Bias-Variance Tradeoff

import pandas as pd
import numpy as np
import seaborn as sns
%matplotlib inline
```

I will be using a module, called `seaborn`, to visualize data points as a scatter plot and also to graph linear (and higher polynomial) regression models:

```
# ## Brain and body weight

'''
This is a [dataset]) of the average
weight of the body and the brain for
62 mammal species. Let's read it into pandas and
take a quick look:
'''
```

```
df = pd.read_table('http://people.sc.fsu.edu/~jburkardt/
datasets/regression/x01.txt', sep='\s+', skiprows=33,
names=['id','brain','body'], index_col='id')
df.head()
```

	brain	body
id		
1	3.385	44.5
2	0.480	15.5
3	1.350	8.1
4	465.000	423.0
5	36.330	119.5

We are going to take a small subset of the samples to exacerbate the visual representations of bias and variance, as follows:

```
# We're going to focus on a smaller subset in which the body weight is
less than 200:
df = df[df.body < 200]
df.shape
(51, 2)
```

We're actually going to pretend that there are only `51` mammal species in existence. In other words, we are pretending that this is the entire dataset of brain and body weights for every known mammal species.

```
# Let's create a scatterplot
sns.lmplot(x='body', y='brain', data=df, ci=None, fit_reg=False)
sns.plt.xlim(-10, 200)
sns.plt.ylim(-10, 250)
```

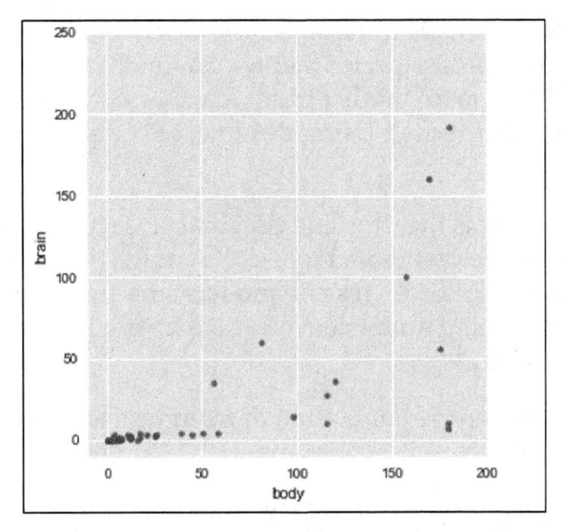

Scatter plot of mammalian brain and body weights

There appears to be a relationship between brain and body weight for mammals. So far, we might assume that it is a positive correlation.

Now, let's throw in a linear regression into the mix. Let's use `seaborn` to make and plot a first degree polynomial (linear) regression.

```
sns.lmplot(x='body', y='brain', data=df, ci=None)
sns.plt.xlim(-10, 200)
sns.plt.ylim(-10, 250)
```

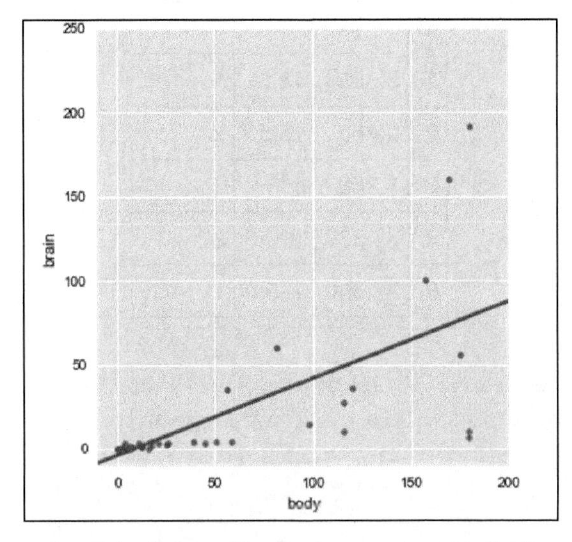

Same scatter plot as before with a linear regression visualization put in

Now, let's pretend that a new mammal species is discovered. We measure the body weight of every member of this species that we can find, and calculate an average body weight of 100. We want to predict the average brain weight of this species (rather than measuring it directly). Using this line, we might predict a brain weight of about 45.

Something you might note is that this line isn't that close to the data points in the graph, so, maybe it isn't the best model to use! You might argue that the *bias is too high*. And I would agree! Linear regression models tend to have high bias, but linear regression also has something up its sleeve—it has a very low variance. However, what does that really mean?

Let's say that we take our entire population of mammals and randomly split them into two samples, as follows:

```
# set a random seed for reproducibility
np.random.seed(12345)


# randomly assign every row to either sample 1 or sample 2
df['sample'] = np.random.randint(1, 3, len(df))
df.head()
```

Include a new sample column:

	brain	body	sample
id			
1	3.385	44.5	1
2	0.480	15.5	2
3	1.350	8.1	2
5	36.330	119.5	2
6	27.660	115.0	1

```
# Compare the two samples, they are fairly different!
df.groupby('sample')[['brain', 'body']].mean()
```

	brain	body
sample		
1	18.113778	52.068889
2	13.323364	34.669091

We can now tell `seaborn` to create two plots, in which the left plot only uses the data from sample 1 and the right plot only uses the data from sample 2:

```
# col='sample' subsets the data by sample and creates two
# separate plots
sns.lmplot(x='body', y='brain', data=df, ci=None, col='sample')
sns.plt.xlim(-10, 200)
sns.plt.ylim(-10, 250)
```

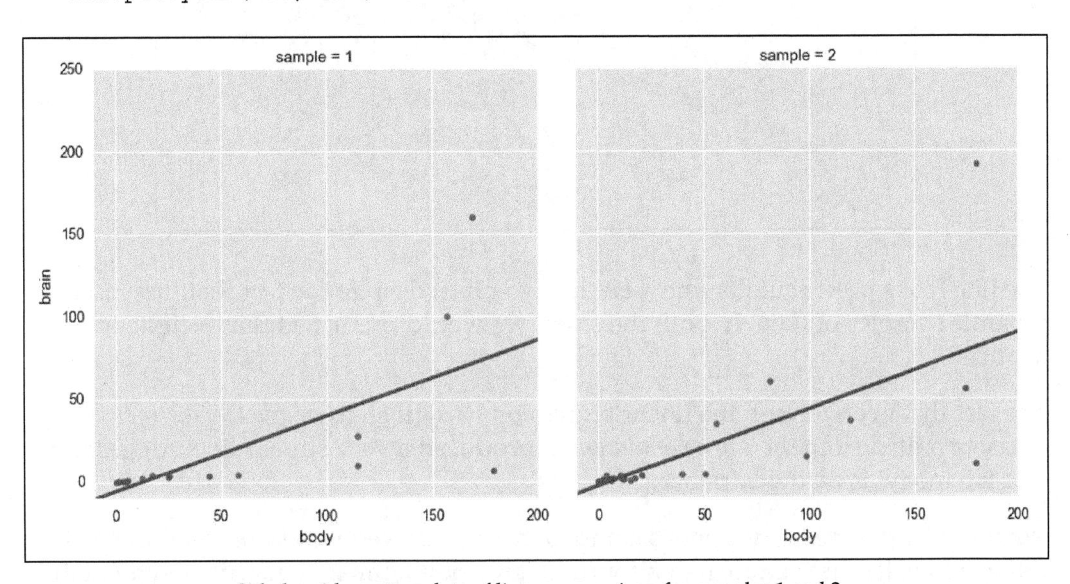

Side-by-side scatter plots of linear regressions for samples 1 and 2

They barely look different, right? If you look closely, you will note that not a single data point is shared between the samples and yet the line looks almost identical. To further show this point, let's put both the lines of best fit in the same graph and use colors to separate the samples, as illustrated:

```
# hue='sample' subsets the data by sample and creates a
# single plot
```

```
sns.lmplot(x='body', y='brain', data=df, ci=None, hue='sample')
sns.plt.xlim(-10, 200)
sns.plt.ylim(-10, 250)
```

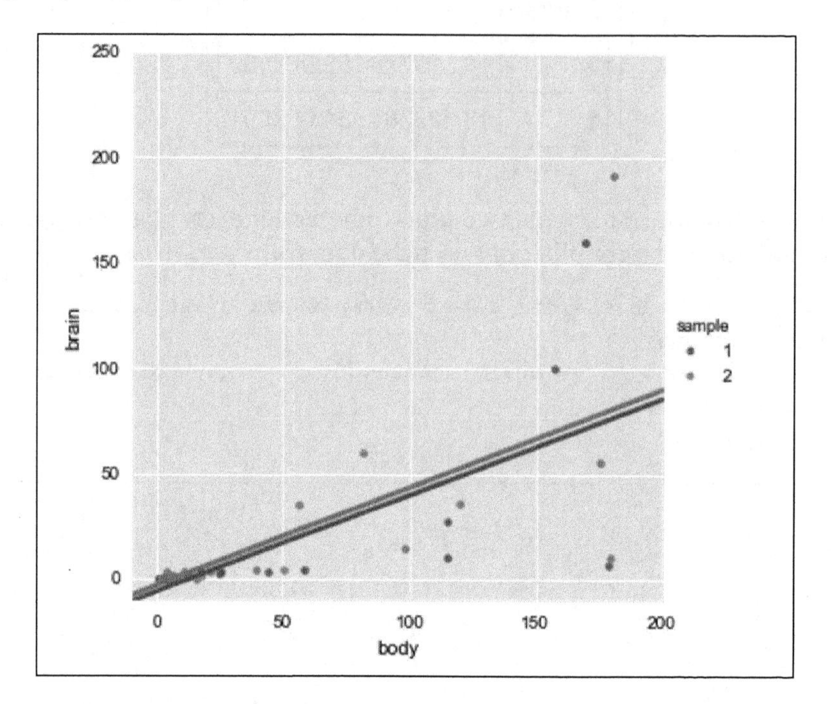

The line looks pretty similar between the two plots, despite the fact that they used separate samples of data. In both the cases, we would predict a brain weight of about 45.

The fact that even though the linear regression was given to *completely distinct* datasets pulled from the *same population*, it produced a very similar line, suggesting that the model is of low variance.

What if we increased our model's complexity and allowed it to *learn more*? Instead of fitting a line, let's let `seaborn` fit a fourth degree polynomial (a quartic polynomial). By adding to the degree of the polynomial, the graph will be able to make twists and turns in order to fit our data better, as shown:

```
# What would a low bias, high variance model look like? Let's try
polynomial regression, with an fourth order polynomial:
sns.lmplot(x='body', y='brain', data=df, ci=None, \
col='sample', order=4)
sns.plt.xlim(-10, 200)
sns.plt.ylim(-10, 250)
```

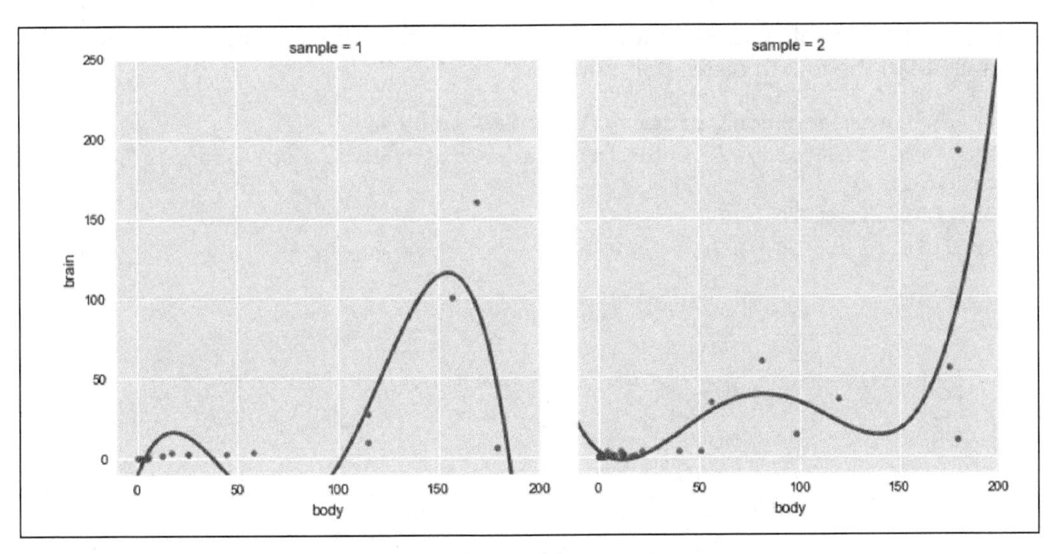

Using a quartic polynomial for regression purposes

Note how, for two distinct samples from the same population, the quartic polynomial looks vastly different. This is a sign of high variance.

This model is low bias because it matches our data well! However, it's high variance because the models are widely different depending upon which points happen to be in the sample. (For a body weight of 100, the brain weight prediction would either be 40 or 0, depending upon which data happened to be in the sample.)

Our polynomial is also unaware of the general relationship of the data. It seems obvious that there is a positive correlation between the brain and body weight of mammals. However, in our quartic polynomials, this relationship is nowhere to be found and is unreliable. In our first sample (the graph on the left), the polynomial ends up shooting downwards, while in the second graph, the graph is going upwards towards the end. Our model is unpredictable and can behave wildly different depending on the given training set.

It is our job, as data scientists, to find a middle ground.

Perhaps we can create a model that has less bias than the linear model, and less variance than the fourth order polynomial?

```
# Let's try a second order polynomial instead:
sns.lmplot(x='body', y='brain', data=df, ci=None, col='sample',
order=2)
sns.plt.xlim(-10, 200)
sns.plt.ylim(-10, 250)
```

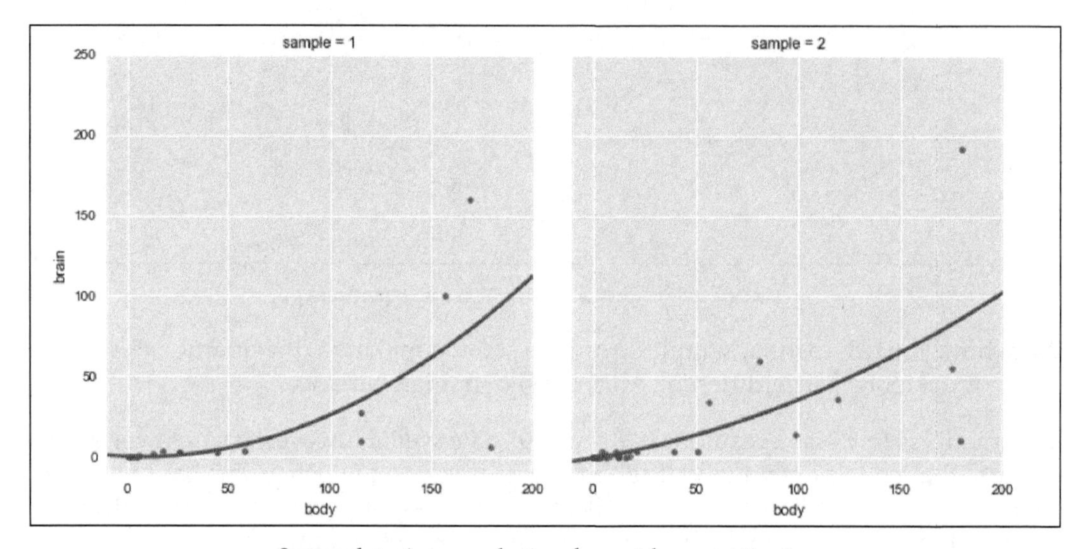

Scatter plot using a quadratic polynomial as our estimator

This plot seems to have a good balance of bias and variance.

Two extreme cases of bias/variance tradeoff

What we just saw were two extreme cases of model fitting: one was underfitting and the other was overfitting.

Underfitting

Underfitting occurs when our models make little to no attempt to fit our data. Models that are high bias and low variance are prone to underfitting. In the case of the mammal brain/body weight example, the linear regression is underfitting our data. While we have a general shape of the relationship, we are left with a high bias.

If your learning algorithm shows high bias and/or is underfitting, the following suggestions may help:

- **Use more features**: Try including new features into the model if it helps with our predictive power.
- **Try a more complicated model**: Adding complexity to your model can help improve bias. An overly complicated model will hurt too!

Overfitting

Overfitting is the result of the model trying too hard to fit into the training set, resulting in a lower bias but a much higher variance. Models that are low bias and high variance are prone to overfitting. In the case of the mammal brain/body weight example, the fourth degree polynomial (quartic) regression is overfitting our data.

If your learning algorithm shows high variance and/or is overfitting, the following suggestions may help:

- **Use fewer features**: Using fewer features can decrease our variance and prevent overfitting
- **Fit on more training samples**: Using more training data points in our cross-validation can reduce the effect of overfitting, and improve our high variance estimator

How bias/variance play into error functions

Error functions (that measure how incorrect our models are) can be thought of as functions of bias, variance, and irreducible error. Mathematically put, the error of predicting a dataset using our supervised learning model might look as follows:

$$Error(x) = Bias^2 + Variance + Irreducible\,Error$$

Here, $Bias^2$ is our bias term squared (arises when simplifying the mentioned statement from more complicated equations), *Variance* is a measurement of how much our model fitting varies between randomized samples.

Simply put, both bias and variance contribute to errors. As we increase our model complexity (for example, go from a linear regression to an eighth degree polynomial regression or grow our decision trees deeper), we find that **Bias²** decreases, variance increases, and the total error of the model forms a parabolic shape, as illustrated:

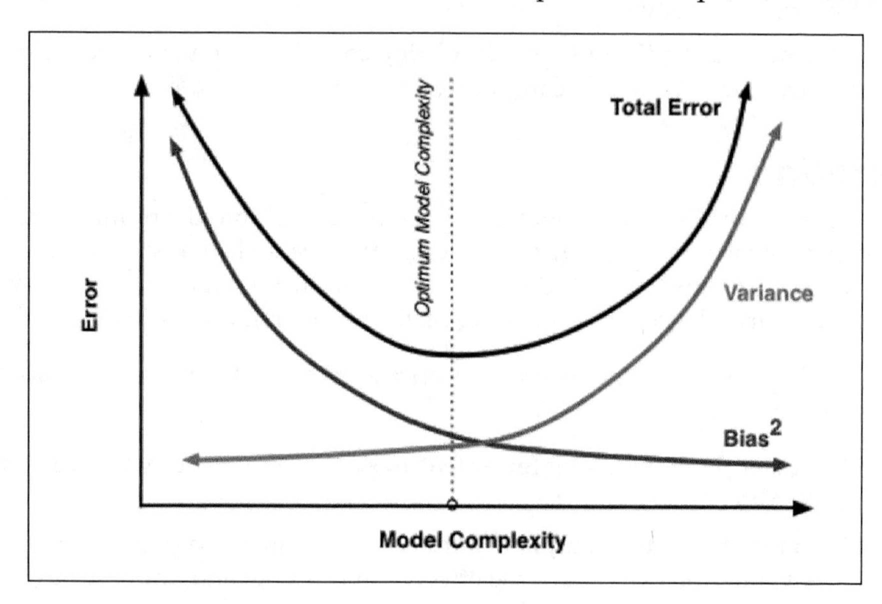

Our goal, as data scientists, is to find the *sweet spot* that optimizes our model complexity. It is easy to overfit our data. To combat overfitting in practice, we should always use cross-validation (splitting up datasets iteratively and retraining models and averaging metrics) to get the best predictor of an error.

To illustrate this point, I will introduce (quickly) a new supervised algorithm and demonstrate the bias/variance tradeoff visually.

We will be using the **K-Nearest Neighbors** (**KNN**) algorithm, which is a supervised learning algorithm that uses a lookalike paradigm, which means that it makes predictions based on similar data points seen in the past.

KNN has a complexity input, K, which represents how many similar data points to compare to. If $K = 3$, then, for a given input, we look to the nearest three data points and use them for our prediction. In this case, K represents our model complexity.

```
from sklearn.neighbors import KNeighborsClassifier
# read in the iris data
from sklearn.datasets import load_iris
iris = load_iris()
X, y = iris.data, iris.target
```

So, we have our X and our y. A great way to overfit a model is to train and predict on the exact same data.

```
knn = KNeighborsClassifier(n_neighbors=1)
knn.fit(X, y)
knn.score(X, y)
1.0
```

Wow, a 100% accuracy?! This is too good to be true.

By training and predicting on the same data, we are essentially telling our data to purely memorize the training set and spit it back to us (this is called our training error). This is the reason we introduced our training and test sets in *Chapter 10, How to Tell If Your Toaster Is Learning – Machine Learning Essentials.*

K folds cross-validation

K folds cross-validation is a much better estimator of our model's performance, even more so than our train-test split. Here's how it works:

1. We will take a finite number of equal slices of our data (usually 3, 5, or 10). Assume that this number is called k.
2. For each "fold" of the cross-validation, we will treat k-1 of the sections as the training set, and the remaining section as our test set.
3. For the remaining folds, a different arrangement of k-1 sections is considered for our training set and a different section is our training set.
4. We compute a set metric for each fold of the cross-validation.
5. We average our scores at the end.

Cross-validation is effectively using multiple train-test splits being done on the same dataset. This is done for a few reasons, but mainly because cross-validation is the most honest estimate of our model's out of the sample error.

To explain this visually, let's look at our mammal brain and body weight example for a second. The following code manually creates a five-fold cross-validation, wherein five different training and test sets are made from the same population:

```
from sklearn.cross_validation import KFold

df = pd.read_table('http://people.sc.fsu.edu/~jburkardt/
datasets/regression/x01.txt', sep='\s+', skiprows=33,
names=['id','brain','body'])
df = df[df.brain < 300][df.body < 500]
```

```
# limit points for visibility

nfolds = 5
fig, axes = plt.subplots(1, nfolds, figsize=(14,4))
for i, fold in enumerate(KFold(len(df), n_folds=nfolds,
                                 shuffle=True)):
    training, validation = fold
    x, y = df.iloc[training]['body'], df.iloc[training]['brain']
    axes[i].plot(x, y, 'ro')
    x, y = df.iloc[validation]['body'], df.iloc[validation]['brain']
    axes[i].plot(x, y, 'bo')
plt.tight_layout()
```

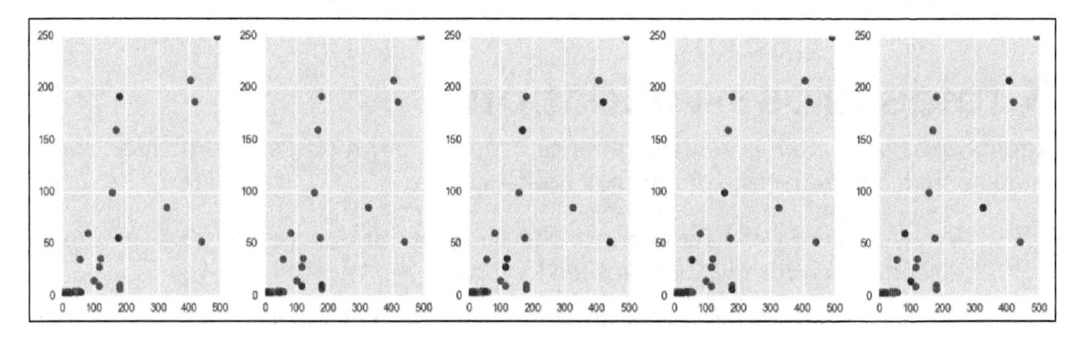

Five-fold cross-validation: red = training sets, blue = test sets

Here, each graph shows the exact same population of mammals, but the dots are colored red if they belong to the training set of that fold and blue if they belong to the testing set. By doing this, we are obtaining five different instances of the same machine learning model in order to see if performance remains consistent across the folds.

If you stare at the dots long enough, you will note that each dot appears in a training set exactly four times (k – 1), while the same dot appears in a test set exactly once and only once.

Some features of K-fold cross-validation include the following:

- It is a more accurate estimate of the OOS prediction error than a single train-test split because it is taking several independent train-test splits and averaging the results together.
- It is a more efficient use of data than single train-test splits because the entire dataset is being used for multiple train-test splits instead of just one.
- Each record in our dataset is used for both training and testing.

- This method presents a clear tradeoff between efficiency and computational expense. A 10-fold CV is *10x* more expensive computationally than a single train/test split.
- This method can be used for parameter tuning and model selection.

Basically, whenever we wish to test a model on a set of data, whether we just completed tuning some parameters or feature engineering, a k-fold cross-validation is an excellent way to estimate the performance on our model.

Of course, `sklearn` comes with an easier-to-use cross-validation module, called `cross_val_score`, which automatically splits up our dataset for us, runs the model on each fold, and gives us a neat and tidy output of results:

```
# Using a training set and test set is so important
# Just as important is cross validation. Remember cross validation
# is using several different train test splits and
# averaging your results!

## CROSS-VALIDATION

# check CV score for K=1
from sklearn.cross_validation import cross_val_score, train_test_split
tree = KNeighborsClassifier(n_neighbors=1)
scores = cross_val_score(tree, X, y, cv=5, scoring='accuracy')
scores.mean()
0.95999999999
```

Which is a much more reasonable accuracy than our previous score of `1`. Remember that we are not getting 100% accuracy anymore, because we have a distinct training and test set. The data points that KNN has never seen the test points and therefore cannot match them exactly to themselves.

Let's try cross-validating KNN with `K=5` (increasing our model's complexity), as shown:

```
# check CV score for K=5
knn = KNeighborsClassifier(n_neighbors=5)
scores = cross_val_score(knn, X, y, cv=5, scoring='accuracy')
scores
np.mean(scores)
0.97333333
```

Even better! So, now we have to find the *best* K? The best K is the one that maximizes our accuracy. Let's try a few:

```
# search for an optimal value of K
k_range = range(1, 30, 2) # [1, 3, 5, 7, …, 27, 29]
errors = []
for k in k_range:
    knn = KNeighborsClassifier(n_neighbors=k)
   # instantiate a KNN with k neighbors
   scores = cross_val_score(knn, X, y, cv=5, scoring='accuracy')
 # get our five accuracy scores
    accuracy = np.mean(scores)
   # average them together
    error = 1 - accuracy
   # get our error, which is 1 minus the accuracy
    errors.append(error)
   # keep track of a list of errors
```

We now have an error value (1 – accuracy) for each value of K (1, 3, 5, 7, 9.., .., 29):

```
# plot the K values (x-axis) versus the 5-fold CV score (y-axis)
plt.figure()
plt.plot(k_range, errors)
plt.xlabel('K')
plt.ylabel('Error')
```

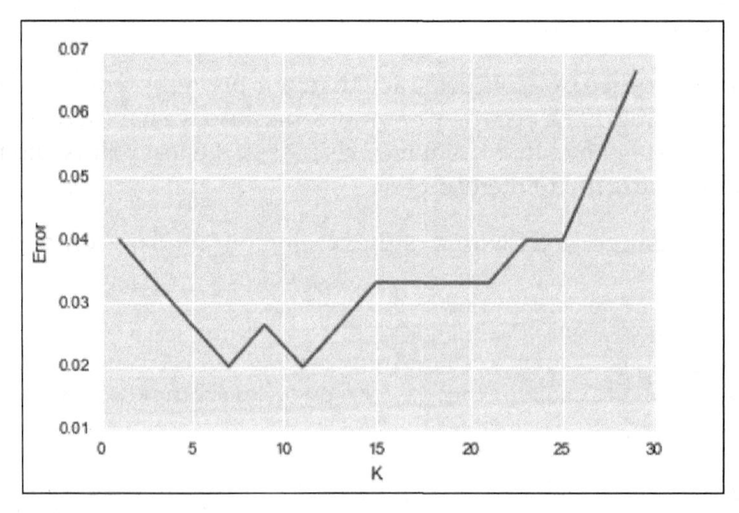

Graph of errors of KNN model against KNN's complexity, represented by the value of K

Compare this graph to the previous graph of model complexity and bias/variance. Toward the left, our graph has a higher bias and is underfitting. As we increased our model's complexity, the error term began to go down, but after a while, our model became overly complex, and the high variance kicked in, making our error term go back up.

It seems that the optimal value of K is between 6 and 10.

Grid searching

`sklearn` also has, up its sleeve, another useful tool called grid searching. A grid search will by brute force try many different model parameters and give us the best one based on a metric of our choosing. For example, we can choose to optimize KNN for accuracy in the following manner:

```
from sklearn.grid_search import GridSearchCV
# import our grid search module

knn = KNeighborsClassifier()
# instantiate a blank slate KNN, no neighbors

k_range = range(1, 30, 2)
param_grid = dict(n_neighbors=k_range)
# param_grid = {"n_ neighbors": [1, 3, 5, …]}

grid = GridSearchCV(knn, param_grid, cv=5, scoring='accuracy')

grid.fit(X, y)
```

In the `grid.fit()` line of code, what is happening is that, for each combination of features, in this case we have 15 different possibilities for K, we are cross-validating each one five times. This means that by the end of this code, we will have *15 * 5 = 75* different KNN models! You can see how, when applying this technique to more complex models, we could run into difficulties with time:

```
# check the results of the grid search
grid.grid_scores_
grid_mean_scores = [result[1] for result in grid.grid_scores_]
# this is a list of the average accuracies for each parameter
# combination
plt.figure()
plt.ylim([0.9, 1])
plt.xlabel('Tuning Parameter: N nearest neighbors')
plt.ylabel('Classification Accuracy')
plt.plot(k_range, grid_mean_scores)
```

```
plt.plot(grid.best_params_['n_neighbors'], grid.best_score_, 'ro',
markersize=12, markeredgewidth=1.5,
         markerfacecolor='None', markeredgecolor='r')
```

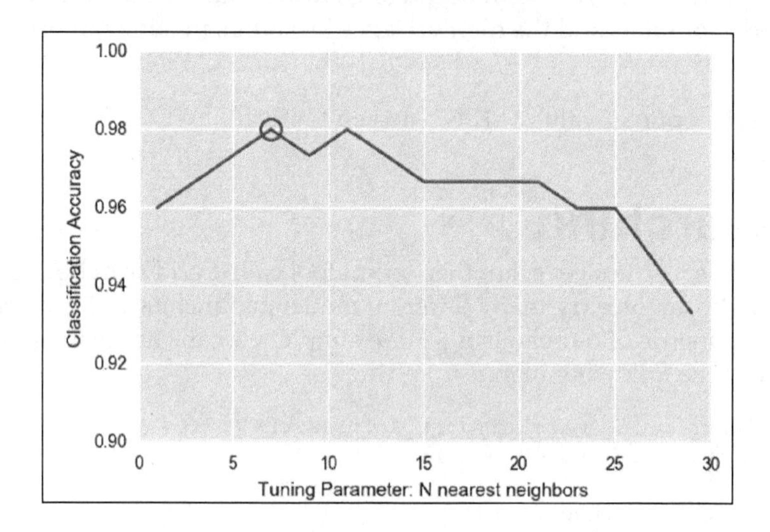

Note that the preceding graph is basically the same as the one we achieved previously with our `for` loop, but much easier!

We see that seven neighbors (circled in the preceding graph) seem to have the best accuracy. However, we can also, very easily, get our best parameters and our best model, as shown:

```
grid.best_params_
# {'n_neighbors': 7}

grid.best_score_
# 0.9799999999

grid.best_estimator_
# actually returns the unfit model with the best parameters
# KNeighborsClassifier(algorithm='auto', leaf_size=30,
metric='minkowski',
           metric_params=None, n_jobs=1, n_neighbors=7, p=2,
           weights='uniform')
```

I'll take this one step further. Maybe you've noted that KNN has other parameters as well, such as `algorithm`, `p`, and `weights`. A quick look at the scikit-learn documentation reveals that we have some options for each of these, which are as follows:

- `p` is an integer and represents the type of distance we wish to use. By default, we use `p=2`, which is our standard distance formula.
- Weights is, by default, `uniform`, but can also be `distance`, which weighs points by their distance, which means that closer neighbors have a greater impact on the prediction.
- Algorithm is how the model finds the nearest neighbors. We can try `ball_tree`, `kd_tree`, or `brute`. The default is `auto`, which tries to use the best one automatically.

```
knn = KNeighborsClassifier()
k_range = range(1, 30)
algorithm_options = ['kd_tree', 'ball_tree', 'auto', 'brute']
p_range = range(1, 8)
weight_range = ['uniform', 'distance']
param_grid = dict(n_neighbors=k_range, weights=weight_range,
algorithm=algorithm_options, p=p_range)
# trying many more options
grid = GridSearchCV(knn, param_grid, cv=5, scoring='accuracy')
grid.fit(X, y)
```

The preceding code takes about a minute to run on my laptop because it is trying many, 1, 648, different combinations of parameters and cross-validating each one five times. All in all, to get the best answer, it is fitting 8,400 different KNN models!

```
grid.best_score_
0.98666666

grid.best_params_
{'algorithm': 'kd_tree', 'n_neighbors': 6, 'p': 3, 'weights':
'uniform'}
```

Grid searching is a simple (but inefficient) way of parameter tuning our models to get the best possible outcome. It should be noted that to get the best possible outcome, data scientists should use feature manipulation (both reduction and engineering) to obtain better results in practice as well. It should not merely be up to the model to achieve the best performance.

Visualizing training error versus cross-validation error

I think it is important once again to go over and compare the cross-validation error and the training error. This time, let's put them both on the same graph to compare how they both change as we vary the model complexity.

I will use the mammal dataset once more to show the cross-validation error and the training error (the error on predicting the training set). Recall that we are attempting to regress the body weight of a mammal to the brain weight of a mammal.

```
# This function uses a numpy polynomial fit function to
# calculate the RMSE of given X and y
def rmse(x, y, coefs):
    yfit = np.polyval(coefs, x)
    rmse = np.sqrt(np.mean((y - yfit) ** 2))
    return rmse

xtrain, xtest, ytrain, ytest = train_test_split(df['body'],
df['brain'])

train_err = []
validation_err = []
degrees = range(1, 8)

for i, d in enumerate(degrees):
    p = np.polyfit(xtrain, ytrain, d)
  # built in numpy polynomial fit function

    train_err.append(rmse(xtrain, ytrain, p))
    validation_err.append(rmse(xtest, ytest, p))


fig, ax = plt.subplots()
# begin to make our graph

ax.plot(degrees, validation_err, lw=2, label = 'cross-validation
error')
ax.plot(degrees, train_err, lw=2, label = 'training error')
# Our two curves, one for training error, the other for cross
validation

ax.legend(loc=0)
ax.set_xlabel('degree of polynomial')
ax.set_ylabel('RMSE')
```

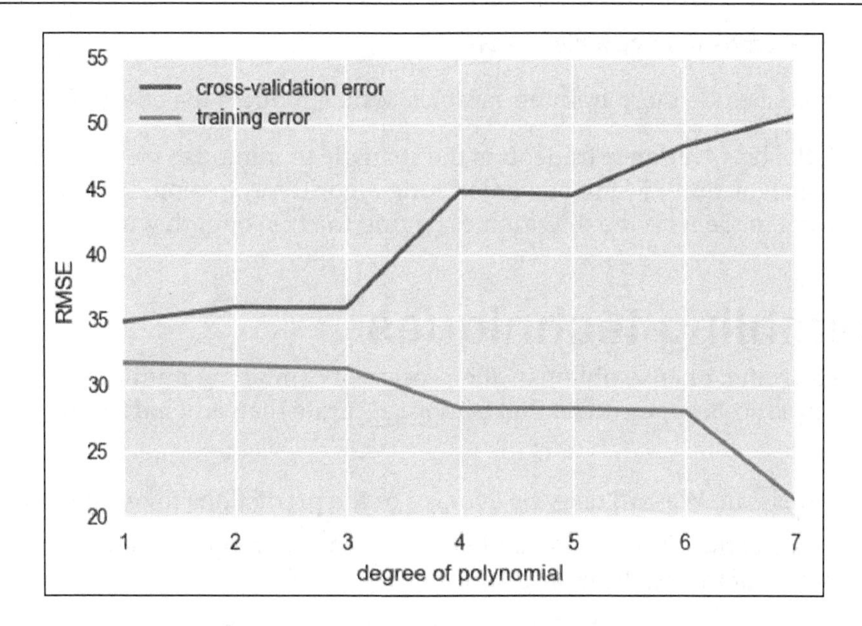

So, we see that as we increase our degree of fit, our training error goes down without a hitch, but we are now smart enough to know that as we increase the model complexity, our model is overfitting to our data and is merely regurgitating our data back to us, whereas our cross validation error line is much more honest and begins to perform poorly after about degree 2 or 3.

To recap:

- Underfitting occurs when the cross-validation error and the training error are both high
- Overrfitting occurs when the cross-validation error is high, while the training error is low
- We have a good fit when the cross-validation error is low, and only slightly higher than the training error

Both underfitting (high bias) and overfitting (high variance) will result in poor generalization of the data.

Here are some tips if you face high bias or variance.

If your model tends to have a *high bias*:

- Try adding more features to the training and test sets
- Either add to the complexity of your model or try a more modern sophisticated model

If your model tends to have a *high variance*:

- Try to include more training samples, which reduces the effect of overfitting

In general, the bias/variance tradeoff is the struggle to minimize bias and variance in our learning algorithms. Many newer learning algorithms, invented in the past few decades, were made with the intention of having the best of both worlds.

Ensembling techniques

Ensemble learning, or ensembling, is the process of combining multiple predictive models to produce a supermodel that is more accurate than any individual model on its own.

- **Regression**: We will take the average of the predictions for each model
- **Classification**: Take a vote and use the most common prediction, or take the average of the predicted probabilities

Imagine that we are working on a binary classification problem (predicting either 0 or 1).

```
# ENSEMBLING

import numpy as np

# set a seed for reproducibility
np.random.seed(12345)

# generate 1000 random numbers (between 0 and 1) for each model,
representing 1000 observations
mod1 = np.random.rand(1000)
mod2 = np.random.rand(1000)
mod3 = np.random.rand(1000)
mod4 = np.random.rand(1000)
mod5 = np.random.rand(1000)
```

Now, we simulate five different learning models that each have about a 70% accuracy, as follows:

```
# each model independently predicts 1 (the "correct response") if
random number was at least 0.3
preds1 = np.where(mod1 > 0.3, 1, 0)
preds2 = np.where(mod2 > 0.3, 1, 0)
preds3 = np.where(mod3 > 0.3, 1, 0)
preds4 = np.where(mod4 > 0.3, 1, 0)
```

```
preds5 = np.where(mod5 > 0.3, 1, 0)

print preds1.mean()
0.699
print preds2.mean()
0.698
print preds3.mean()
0.71
print preds4.mean()
0.699
print preds5.mean()
0.685

# Each model has an "accuracy of around 70% on its own
```

Now, let's apply my degrees in magic. Er sorry, math.

```
# average the predictions and then round to 0 or 1
ensemble_preds = np.round((preds1 + preds2 + preds3 + preds4 +
preds5)/5.0).astype(int)
ensemble_preds.mean()

0.83
```

As you add more models to a voting process, the probability of errors will decrease; this is known as Condorcet's jury theorem.

Crazy, right?

For ensembling to work well in practice, the models must have the following characteristics:

- **Accuracy**: Each model must at least outperform the null model
- **Independence**: A model's prediction is not affected by another model's prediction process

If you have a bunch of individually *OK* models, the edge case mistakes made by one model are probably not going to be made by the other models, so the mistakes will be ignored when combining the models.

There are the following two basic methods for ensembling:

- Manually ensemble your individual models by writing a good deal of code
- Use a model that ensembles for you

We're going to look at a model that ensembles for us. To do this, let's take a look at decision trees again.

Decision trees tend to have low bias and high variance. Given any dataset, the tree can keep asking questions (making decisions) until it is able to nitpick and distinguish between *every single* example in the dataset. It could keep asking question after question until there is only a single example in each leaf (terminal) node. The tree is trying too hard, growing too deep, and just memorizing every single detail of our training set. However, if we started over, the tree could potentially ask different questions and still grow very deep. This means that there are many possible trees that could distinguish between all elements, which means higher variance. It is unable to generalize well.

In order to reduce the variance of a single tree, we can place a restriction on the number of questions asked in a tree (the `max_depth` parameter) or we can create an ensemble version of decision trees, called Random forests.

Random forests

The primary weakness of decision trees is that different splits in the training data can lead to very different trees. Bagging is a general purpose procedure to reduce the variance of a machine learning method, but is particularly useful for decision trees.

Bagging is short for Bootstrap aggregation, which means the aggregation of Bootstrap samples. What is a Bootstrap sample? It is a random sample with replacement:

```
# set a seed for reproducibility
np.random.seed(1)

# create an array of 1 through 20
nums = np.arange(1, 21)
print nums
[ 1  2  3  4  5  6  7  8  9 10 11 12 13 14 15 16 17 18 19 20]


# sample that array 20 times with replacement
np.random.choice(a=nums, size=20, replace=True)
[ 6 12 13  9 10 12  6 16  1 17  2 13  8 14  7 19  6 19 12 11]
# This is our bootstrapped sample notice it has repeat variables!
```

So, how does bagging work for decision trees?

1. Grow B trees using Bootstrap samples from the training data.
2. Train each tree on its Bootstrap sample and make predictions.
3. Combine the predictions:
    - Average the predictions for regression trees
    - Take a vote for classification trees

The following are a few things to note:

- Each Bootstrap sample should be the same size as the original training set
- B should be a large enough value that the error seems to have *stabilized*
- The trees are grown intentionally deep so that they have low bias/high variance

The reason we grow the trees intentionally deep is because the bagging inherently increases predictive accuracy by reducing the variance, similar to how cross-validation reduces the variance associated with estimating our out of sample error.

Random forests are a variation of bagged trees.

However, when building each tree, each time we consider a split between the features, a random sample of *m* features is chosen as split candidates from the full set of `p` features. The split is only allowed to be one of those `m` features:

- A new random sample of features is chosen for *every single* tree at every single split
- For classification, *m* is typically chosen to be the square root of `p`
- For regression, `m` is typically chosen to be somewhere between `p/3` and `p`

What's the point?

Suppose there is one very strong feature in the dataset. When using decision (or bagged) trees, most of the trees will use that feature as the top split, resulting in an ensemble of similar trees that are highly correlated to each other.

If our trees are highly correlated to each other, then averaging these quantities will not significantly reduce variance (which is the entire goal of ensembling). Also, by randomly leaving out candidate features from each split, Random forests reduce the variance of the resulting model.

Random forests can be used in both classification and regression problems and can be easily used in scikit-learn. Let's try to predict MLB salaries based on statistics about the player, as shown:

```
# read in the data
url = '../data/hitters.csv'

hitters = pd.read_csv(url)

# remove rows with missing values
hitters.dropna(inplace=True)

# encode categorical variables as integers
hitters['League'] = pd.factorize(hitters.League)[0]
hitters['Division'] = pd.factorize(hitters.Division)[0]
hitters['NewLeague'] = pd.factorize(hitters.NewLeague)[0]

# define features: exclude career statistics (which start with "C")
and the response (Salary)
feature_cols = [h for h in hitters.columns if h[0] != 'C' and h !=
'Salary']



# define X and y
X = hitters[feature_cols]
y = hitters.Salary
```

Let's try and predict the salary first using a single decision tree, as illustrated:

```
from sklearn.tree import DecisionTreeRegressor

# list of values to try for max_depth
max_depth_range = range(1, 21)

# list to store the average RMSE for each value of max_depth
RMSE_scores = []

# use 10-fold cross-validation with each value of max_depth
from sklearn.cross_validation import cross_val_score
for depth in max_depth_range:
    treereg = DecisionTreeRegressor(max_depth=depth, random_state=1)
    MSE_scores = cross_val_score(treereg, X, y, cv=10, scoring='mean_
squared_error')
    RMSE_scores.append(np.mean(np.sqrt(-MSE_scores)))
```

```
# plot max_depth (x-axis) versus RMSE (y-axis)
plt.plot(max_depth_range, RMSE_scores)
plt.xlabel('max_depth')
plt.ylabel('RMSE (lower is better)')
```

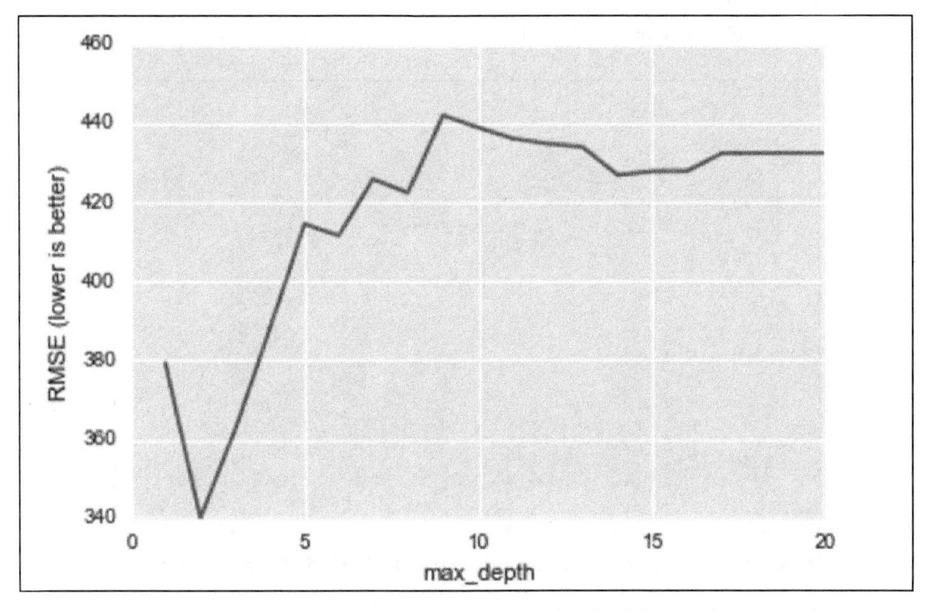

RMSE for decision tree models against the max depth of the tree (complexity)

Let's do the same thing, but this time with a Random forest:

```
from sklearn.ensemble import RandomForestRegressor


# list of values to try for n_estimators
estimator_range = range(10, 310, 10)

# list to store the average RMSE for each value of n_estimators
RMSE_scores = []

# use 5-fold cross-validation with each value of n_estimators
(WARNING: SLOW!)
for estimator in estimator_range:
    rfreg = RandomForestRegressor(n_estimators=estimator, random_
state=1)
    MSE_scores = cross_val_score(rfreg, X, y, cv=5, scoring='mean_
squared_error')
```

```
    RMSE_scores.append(np.mean(np.sqrt(-MSE_scores)))


# plot n_estimators (x-axis) versus RMSE (y-axis)
plt.plot(estimator_range, RMSE_scores)
plt.xlabel('n_estimators')
plt.ylabel('RMSE (lower is better)')
```

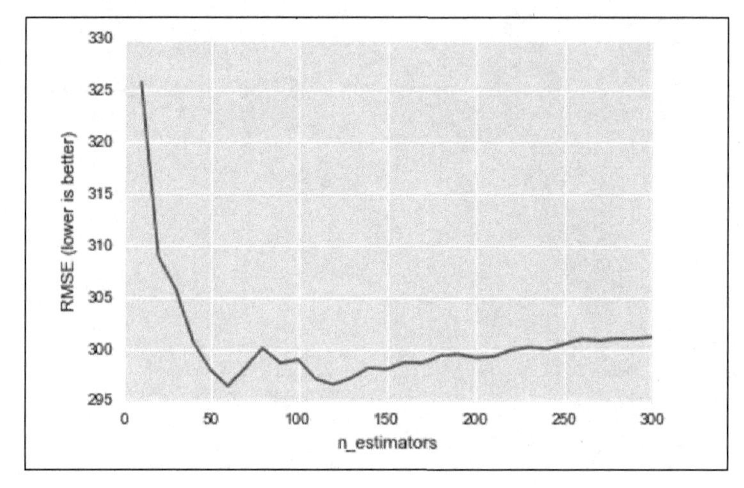

RMSE for random forest models against the max depth of the tree (complexity)

Note already the *y* axis, our RMSE is much lower on an average! See how we can obtain a major increase in predictive power using Random forests.

In Random forests, we still have the concept of important features like we had in decision trees:

```
# n_estimators=150 is sufficiently good
rfreg = RandomForestRegressor(n_estimators=150, random_state=1)
rfreg.fit(X, y)
# compute feature importances
pd.DataFrame({'feature':feature_cols, 'importance':rfreg.feature_
importances_}).sort('importance', ascending = False)
```

	feature	importance
6	Years	0.263990
5	Walks	0.146786
1	Hits	0.139801
4	RBI	0.136265
0	AtBat	0.091551
9	PutOuts	0.060647
3	Runs	0.057460
2	HmRun	0.040183
11	Errors	0.024711
10	Assists	0.023367
8	Division	0.007628
12	NewLeague	0.004545
7	League	0.003067

So, it looks like the number of years the player has been in the league is still the most important feature when deciding that player's salary.

Comparing Random forests with decision trees

It is important to realize that just using random forests is not the solution to your data science problems. While random forests provides many advantages, many disadvantages also come with them, as listed.

The advantages of Random forests are as follows:

- Its performance is competitive with the best supervised learning methods
- It provides a more reliable estimate of feature importance
- It allows you to estimate out-of-sample errors without using train/test splits or cross-validation

The disadvantages of Random forests are as follows:

- It is less interpretable (cannot visualize an entire forest of decision trees)
- It is slower to train and predict (not great for production or real-time purposes

Neural networks

Probably one of the most talked about machine learning models, neural networks are computational networks built to model animals' nervous systems. Before getting too deep into the structure, let's take a look at the big advantages of neural networks.

The key component of neural networks is that it is not only a complex structure, it is a complex and flexible structure. This means the following two things:

- Neural networks are able to estimate any function shape (this is called being non-parametric)
- Neural networks can adapt and literally change their own internal structure based on their environment

Basic structure

Neural networks are made up of interconnected nodes (**perceptrons**) that each take in input (quantitative value), and output other quantitative values. Signals travel through the network and eventually end up at a prediction node.

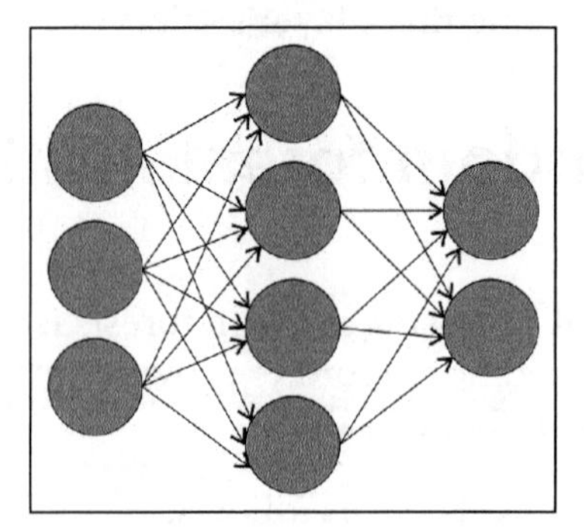

Visualization of neural network interconnected nodes

Another huge advantage of neural networks is that they can be used for supervised learning, unsupervised learning, and reinforcement learning problems. The ability to be so flexible, predict many functional shapes, and adapt to their surroundings make neural networks highly preferable in select fields, as follows:

- **Pattern recognition**: This is probably the most common application of neural networks. Some examples are handwriting recognition and image processing (facial recognition).

- **Entity movement**: Examples for this include self-driving cars, robotic animals, and drone movement.
- **Anomaly detection**: As neural networks are good at recognizing patterns, they can also be used to recognize when a data point does not fit a pattern. Think of a neural network monitoring a stock price movement; after a while of learning the general pattern of a stock price, the network can alert you when something is unusual in the movement.

The simplest form of a neural network is a single perceptron. A perceptron, visualized as follows, takes in some input and outputs a signal:

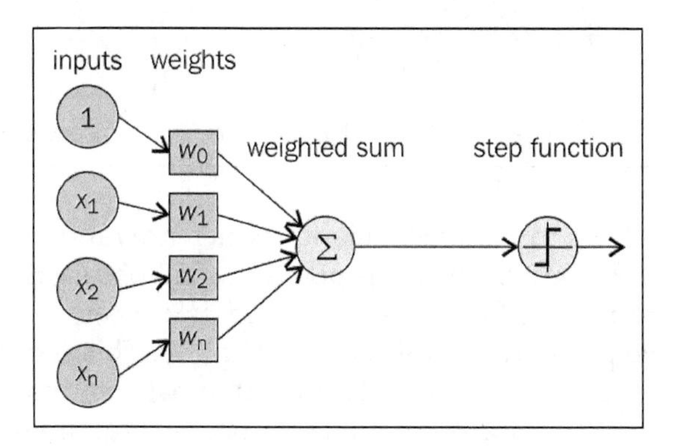

This signal is obtained by combining the input with several weights and then is put through some *activation function*. In cases of simple binary outputs, we generally use the logistic function, as shown:

$$f_{log}(z) = \frac{1}{1 + e^{-z}}$$

f_{log} is called logistic function

To create a neural network, we need to connect multiple perceptrons to each other in a network fashion, as illustrated in the following graph.

A **multilayer perceptrons** (**MLP**) is a finite acyclic graph. The nodes are neurons with logistic activation.

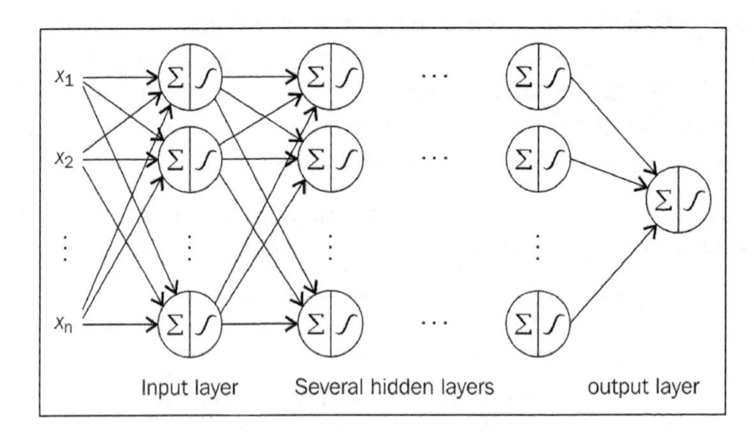

As we train the model, we update the weights (which are random at first) of the model in order to get the best predictions possible. If an observation goes through the model and is outputted as false when it should have been true, the logistic functions in the single perceptrons are changed slightly. This is called **back-propagation**. Neural networks are usually trained in batches, which means that the network is given several training data points at once several times, and each time, the back-propagation algorithm will trigger an internal weight change in the network.

It isn't hard to see that we can grow the network very deep and have many hidden layers, which are associated to the complexity of the neural network. When we grow our neural networks very deep, we are dipping our toes into the idea of *deep learning*. The main advantage of deep neural networks (networks with many layers) is that they can approximate almost any shape function and they can (theoretically) learn optimal combinations of features for us and use these combinations to obtain the best predictive power.

Let's see it in action. I will be using a module, called PyBrain, to make my neural networks. However, first, let's take a look at a new dataset, which is a dataset of handwritten digits. We will first try to recognize digits using a Random forest, as shown:

```
from sklearn.cross_validation import cross_val_score
from sklearn import datasets
import matplotlib.pyplot as plt
from sklearn.ensemble import RandomForestClassifier
%matplotlib inline
```

```
digits = datasets.load_digits()

plt.imshow(digits.images[100], cmap=plt.cm.gray_r,
interpolation='nearest')
# a 4 digit
```

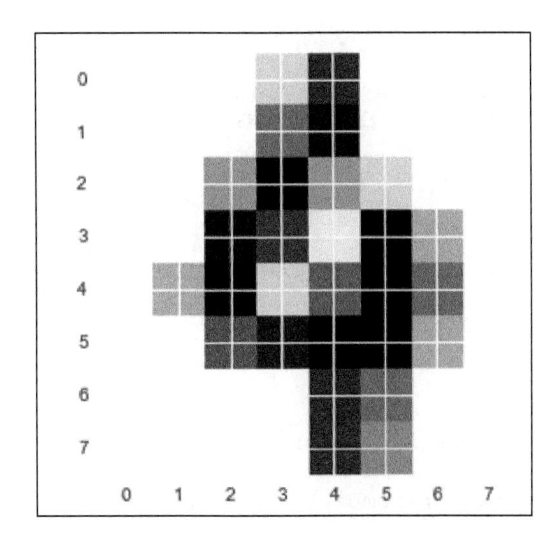


```
X, y = digits.data, digits.target

# 64 pixels per image
X[0].shape

# Try Random Forest
rfclf = RandomForestClassifier(n_estimators=100, random_state=1)
cross_val_score(rfclf, X, y, cv=5, scoring='accuracy').mean()
0.9382782
```

Pretty good! An accuracy of 94% is nothing to laugh at, but can we do even better?

[Warning! The PyBrain syntax can be a bit tricky.]

```
from pybrain.datasets                import ClassificationDataSet
from pybrain.utilities               import percentError
from pybrain.tools.shortcuts         import buildNetwork
from pybrain.supervised.trainers import BackpropTrainer
from pybrain.structure.modules       import SoftmaxLayer
from numpy import ravel
```

```
# pybrain has its own data sample class that we must add
# our training and test set to
ds = ClassificationDataSet(64, 1 , nb_classes=10)
for k in xrange(len(X)):
    ds.addSample(ravel(X[k]),y[k])

# their equivalent of train test split
test_data, training_data = ds.splitWithProportion( 0.25 )




# pybrain's version of dummy variables

test_data._convertToOneOfMany( )
training_data._convertToOneOfMany( )


print test_data.indim # number of pixels going in
# 64
print test_data.outdim # number of possible options (10 digits)
# 10



# instantiate the model with 64 hidden layers (standard params)
fnn = buildNetwork( training_data.indim, 64, training_data.outdim,
outclass=SoftmaxLayer )
trainer = BackpropTrainer( fnn, dataset=training_data, momentum=0.1,
learningrate=0.01 , verbose=True, weightdecay=0.01)

# change the number of epochs to try to get better results!
trainer.trainEpochs (10) # 10 batches
print 'Percent Error on Test dataset: ' , \
        percentError( trainer.testOnClassData (
           dataset=test_data )
           , test_data['class'] )
```

The model will output a final error on a test set:

```
Percent Error on Test dataset: 4.67706013363
accuracy = 1 - .0467706013363
accuracy
0.95322
```

Already better! Both the random forests and neural networks do very well with this problem because both of them are non-parametric, which means that they do not rely on the underlying shape of data to make predictions. They are able to estimate any shape of function.

To predict the shape, we can use the following code:

```
plt.imshow(digits.images[0], cmap=plt.cm.gray_r,
interpolation='nearest')


fnn.activate(X[0])
array([ 0.92183643,  0.00126609,  0.00303146,  0.00387049,
0.01067609,
        0.00718017,  0.00825521,  0.00917995,  0.00696929,
0.02773482])
```

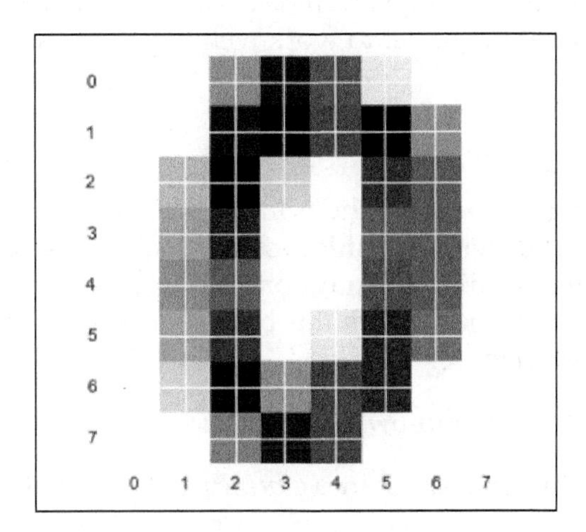

The array represents a probability for every single digit, which means that there is a 92% chance that the digit in the preceding screenshot is a 0 (which it is). Note how the next highest probability is for a 9, which makes sense because 9 and 0 have similar shapes (ovular).

Neural networks do have a major flaw. If left alone, they have a very high variance. To see this, let's run the exact same code as the preceding one and train the exact same type of neural network on the exact same data, as illustrated:

```
# Do it again and see the difference in error
fnn = buildNetwork( training_data.indim, 64, training_data.outdim,
outclass=SoftmaxLayer )
trainer = BackpropTrainer( fnn, dataset=training_data, momentum=0.1,
learningrate=0.01 , verbose=True, weightdecay=0.01)
```

```
# change the number of eopchs to try to get better results!
trainer.trainEpochs (10)
print 'Percent Error on Test dataset: ' , \
        percentError( trainer.testOnClassData (
           dataset=test_data )
           , test_data['class'] )

accuracy = 1 - .0645879732739
accuracy
0.93541
```

See how just rerunning the model and instantiating different weights made the network turn out to be different than before? This is a symptom of being a high variance model. In addition, neural networks generally require many training samples in order to combat the high *varianceness* of the model and also require a large amount of computation power to work well in production environments.

Summary

This concludes our long journey into the principles of data science. In the last 300 odd pages, we looked at different techniques in probability, statistics, and machine learning to answer the most difficult questions out there. I would like to personally congratulate you for making it through this book. I hope that it proved useful and inspired you to learn even more!

This isn't everything I need to know?

Nope! There is only so much I can fit into a *principles* level book. There is still so much to learn.

Where can I learn more?

I recommend going to find open source data challenges (`https://www.kaggle.com/` is a good source) for this. I'd also recommend seeking out, and trying and solving your own problems at home!

When do I get to call myself a data scientist?

When you begin cultivating actionable insights from datasets, both large and small, that companies and people can use, then you have the honor of calling yourself a true data scientist.

13
Case Studies

In this chapter, we will take a look at a few case studies to help you develop a better understanding of the topics we've seen so far.

Case study 1 – predicting stock prices based on social media

Our first case study will be quite exciting! We will attempt to predict the price of stock of a publically traded company using only social media sentiment. While this example will not use any explicit statistical/machine learning algorithms we will utilize EDA (exploratory data analysis) and use visuals in order to achieve our goal.

Text sentiment analysis

When talking about sentiment it should be clear what is meant. By sentiment, I am referring to a quantitative value (at the interval level) between -1 and 1. If the sentiment score of a text piece is close to -1, it is said to have negative sentiment. If the sentiment score is close to 1, then the text is said to have positive sentiment. If the sentiment score is close to 0, we say it has neutral sentiment. We will use a Python module called `Textblob` to measure our text sentiment:

```
# use the textblob module to make a function called stringToSentiment
that returns a sentences sentiment
def stringToSentiment(text):
    return TextBlob(text).sentiment.polarity
```

Now we can use this function that calls the `Textblob` module to score text out of the box:

```
stringToSentiment('i hate you')
-0.8

stringToSentiment('i love you')
0.5

stringToSentiment('i see you')
0.0
```

Now let's read in our tweets for our study:

```
# read in tweets data into a dataframe
from textblob import TextBlob
import pandas as pd
%matplotlib inline
# these tweets are from last May and are about Apple (AAPL)
tweets = pd.read_csv('../data/so_many_tweets.csv')
tweets.head()
```

	Text	Date	Status	Retweet
0	RT @j_o_h_n_danger: $TWTR now top holding for ...	2015-05-24 03:46:08	602319644234395648	6.022899e+17
1	RT diggingplatinum RT WWalkerWW: iOS 9 vs. And...	2015-05-24 04:17:42	602327586983796737	NaN
2	RT bosocial RT insidermonkey RT j_o_h_n_danger...	2015-05-24 04:13:22	602326499534966784	NaN
3	RT @WWalkerWW: iOS 9 vs. Android M â The New...	2015-05-24 04:08:34	602325288740114432	6.023104e+17
4	RT @seeitmarket: Apple Chart Update: Big Test ...	2015-05-24 04:04:42	602324318903771136	6.023215e+17

Exploratory data analysis

So we have four columns:

- `Text`: Unstructured text at the nominal level
- `Date`: Datetime (we will think of datetime in a continuous way)
- `Status`: Status unique ID at the nominal level
- `Retweet`: Status ID of tweet that this tweet was a retweet at the nominal level

So we have four columns, but how many rows? Also what does each row represent? It seems that each row represents a single tweet about the company:

```
tweets.shape
```

```
(52512, 4)
```

So we have four columns and `52512` tweets/rows at our disposal! Oh boy...
Our goal here is to eventually use the tweets' sentiment, so we will likely need a sentiment column in the Dataframe. Using our fairly straightforward function from the previous example, let's add this column!

```
# create a new column in tweets called sentiment that maps
stringToSentiment to the text column
tweets['sentiment'] = tweets['Text'].apply(stringToSentiment)

tweets.head()
```

The preceding code will apply the function `stringToSentiment` to each and every element in the column `Text` of the tweets Dataframe:

```
tweets.head()
```

	Text	Date	Status	Retweet	sentiment
0	RT @j_o_h_n_danger: $TWTR now top holding for ...	2015-05-24 03:46:08	602319644234395648	6.022899e+17	0.500000
1	RT diggingplatinum RT WWalkerWW: iOS 9 vs. And...	2015-05-24 04:17:42	602327586983796737	NaN	0.136364
2	RT bosocial RT insidermonkey RT j_o_h_n_danger...	2015-05-24 04:13:22	602326499534966784	NaN	0.500000
3	RT @WWalkerWW: iOS 9 vs. Android M â The New...	2015-05-24 04:08:34	602325288740114432	6.023104e+17	0.136364
4	RT @seeitmarket: Apple Chart Update: Big Test ...	2015-05-24 04:04:42	602324318903771136	6.023215e+17	0.000000

So now we have a sense for the sentiment score for each tweet in this dataset.
Let's simplify our problem and try to use an entire days worth of tweets to predict whether or not the price of AAPL will increase within 24 hours. If this is the case, we have another issue here. The `Date` column reveals that we have multiple tweets for each day. Just look at the first five tweets; they are all in the same day. We will resample this dataset in order to get a sense of the `average` sentiment of the stock on Twitter every day.

We will do this in three steps:

1. We will ensure that the `Date` column is of the Python `datetime` type.
2. We will replace our Dataframe's index with the datetime column (which allows us to use complex datetime functions).
3. We will resample the data so that each row, instead of representing a tweet, will represent a single day with an aggregated sentiment score for each day:

The index of the Dataframe is a special series used to identify rows in our structure. By default, a Dataframe will use incremental integers to represent rows (`0` for the first row, `1` for the second row, and so on).

```
tweets.index
RangeIndex(start=0, stop=52512, step=1)

# As a list, we can splice it
list(tweets.index)[:5]

[0, 1, 2, 3, 4]
```

4. Let's tackle this date issue now! We will ensure that the `Date` column is of the Python `datetime` type:

```
# cast the date column as a datetime
tweets['Date'] = pd.to_datetime(tweets.Date)
tweets['Date'].head()


Date
2015-05-24 03:46:08   2015-05-24 03:46:08
2015-05-24 04:17:42   2015-05-24 04:17:42
2015-05-24 04:13:22   2015-05-24 04:13:22
2015-05-24 04:08:34   2015-05-24 04:08:34
2015-05-24 04:04:42   2015-05-24 04:04:42

Name: Date, dtype: datetime64[ns]
```

5. We will replace our Dataframe's index with the datetime column (which allows us to use complex datetime functions):

```
tweets.index = tweets.Date
tweets.index
Index([u'2015-05-24 03:46:08', u'2015-05-24 04:17:42', u'2015-05-
24 04:13:22',
       u'2015-05-24 04:08:34', u'2015-05-24 04:04:42', u'2015-05-
24 04:00:01',
       u'2015-05-24 03:54:07', u'2015-05-24 04:25:29', u'2015-05-
24 04:24:47',
       u'2015-05-24 04:06:42',
       ...
       u'2015-05-02 16:30:02', u'2015-05-02 16:29:35', u'2015-05-
02 16:28:26',
       u'2015-05-02 16:27:53', u'2015-05-02 16:27:02', u'2015-05-
02 16:26:39',
       u'2015-05-02 16:25:00', u'2015-05-02 16:23:39', u'2015-05-
02 16:23:38',
       u'2015-05-02 16:23:21'],
      dtype='object', name=u'Date', length=52512)


tweets.head()
```

	yesterday_sentiment	Close	yesterday_close	percent_change_in_price	change_close_big_deal
Date					
2015-05-05	0.084062	125.800003	128.699997	-0.022533	True
2015-05-06	0.063882	125.010002	125.800003	-0.006280	False
2015-05-07	0.066166	125.260002	125.010002	0.002000	False
2015-05-08	0.078892	127.620003	125.260002	0.018841	True
2015-05-11	0.102898	126.320000	127.620003	-0.010187	True

Note that the black index on the left used to be numbers, but now is the exact datetime that the tweet was sent.

6. Resample the data so that each row, instead of representing a tweet, will represent a single day with an aggregated sentiment score for each day:

```
# create a dataframe called daily_tweets which resamples tweets by
D, averaging the columns
daily_tweets = tweets[['sentiment']].resample('D', how='mean')
# I only want the sentiment column in my new Dataframe.
daily_tweets.head()
```

	sentiment
Date	
2015-05-02	0.083031
2015-05-03	0.107789
2015-05-04	0.084062
2015-05-05	0.063882
2015-05-06	0.066166

Now that's looking better! Now each row represents a single day and the sentiment score column is showing us an average sentiment for the day. Let's see how many days worth of tweets we have:

```
daily_tweets.shape
```

```
(23, 3)
```

OK, so we went from over 50,000 tweets to only `23` days! Let's take a look at the progression of sentiment over several days:

```
# plot the sentiment as a line graph
daily_tweets.sentiment.plot(kind='line')
```

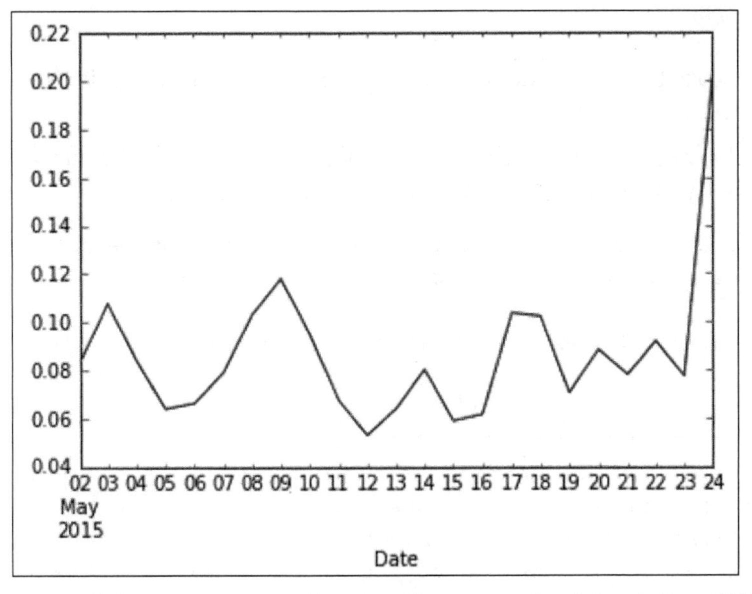

Average daily sentiment in regard to a specific company for 23 days in May 2015

```
# get historical prices through the Yahoo Finance API
from yahoo_finance import Share
yahoo = Share("AAPL")
historical_prices = yahoo.get_historical('2015-05-2', '2015-05-25')
prices = pd.DataFrame(historical_prices)

prices.head()
```

	Adj_Close	Close	Date	High	Low	Open	Symbol	Volume
0	129.180748	132.539993	2015-05-22	132.970001	131.399994	131.600006	AAPL	45596000
1	128.059901	131.389999	2015-05-21	131.630005	129.830002	130.070007	AAPL	39730400
2	126.763608	130.059998	2015-05-20	130.979996	129.339996	130.00	AAPL	36454900
3	126.773364	130.070007	2015-05-19	130.880005	129.639999	130.690002	AAPL	44633200
4	126.890318	130.190002	2015-05-18	130.720001	128.360001	128.380005	AAPL	50882900

Now two things:

- We are really only interested in the `Close` column, which is the final price set for the trading day
- We also need to set the index of this Dataframe to be *datetimes* so that we can merge the sentiment and the price Dataframes together

```
# Set the index of the price dataframe to also be datetimes
prices.index = pd.to_datetime(prices['Date'])

prices.info() #the columns aren't numbers!


<class 'pandas.core.frame.DataFrame'>
DatetimeIndex: 15 entries, 2015-05-22 to 2015-05-04
Data columns (total 8 columns):
Adj_Close     15 non-null object
Close         15 non-null object          # NOT A NUMBER
Date          15 non-null object
High          15 non-null object
Low           15 non-null object
Open          15 non-null object
Symbol        15 non-null object
Volume        15 non-null object
dtypes: object(8)
```

Let's fix that. While we're at it, let's also fix `Volume`, which represents the number of traded stocks in that day:

```
# cast the column as numbers
prices.Close = not_null_close.Close.astype('float')
prices.Volume = not_null_close.Volume.astype('float')
```

Now let's try to plot both the volume and price of AAPL in the same graph:

```
# plot both volume and close as line graphs in the same graph, what do
you notice is the problem?
prices[["Volume", 'Close']].plot()
```

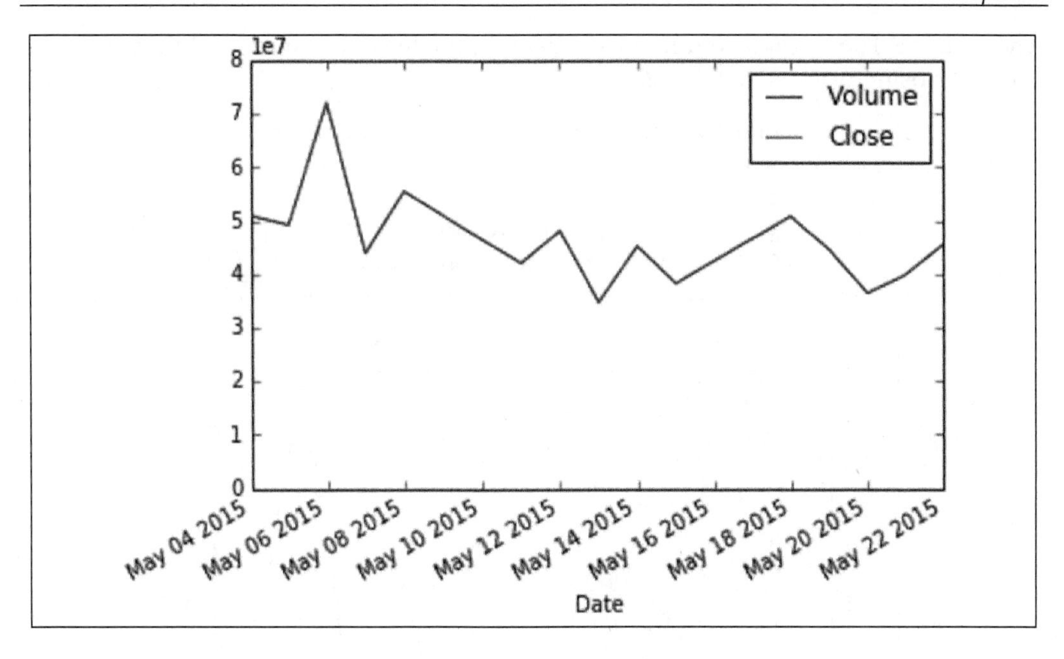

Woah, what's wrong here? Well if we look carefully, `Volume` and `Close` are on very different scales!

```
prices[["Volume", 'Close']].describe()
```

	Volume	Close
count	1.500000e+01	15.000000
mean	4.649939e+07	128.170667
std	9.167054e+06	2.386393
min	3.469420e+07	125.010002
25%	4.088310e+07	125.940002
50%	4.520350e+07	128.699997
75%	5.007715e+07	130.065002
max	7.214100e+07	132.539993

And by a lot! The `Volume` column has a mean in the tens of millions, while the average closing price is merely 125!

```
# scale the columns by z scores using StandardScaler
# Then plot the scaled data
s = StandardScaler()
only_prices_and_volumes = prices[["Volume", 'Close']]
price_volume_scaled = s.fit_transform(only_prices_and_volumes)
pd.DataFrame(price_volume_scaled, columns=["Volume", 'Close']).plot()
```

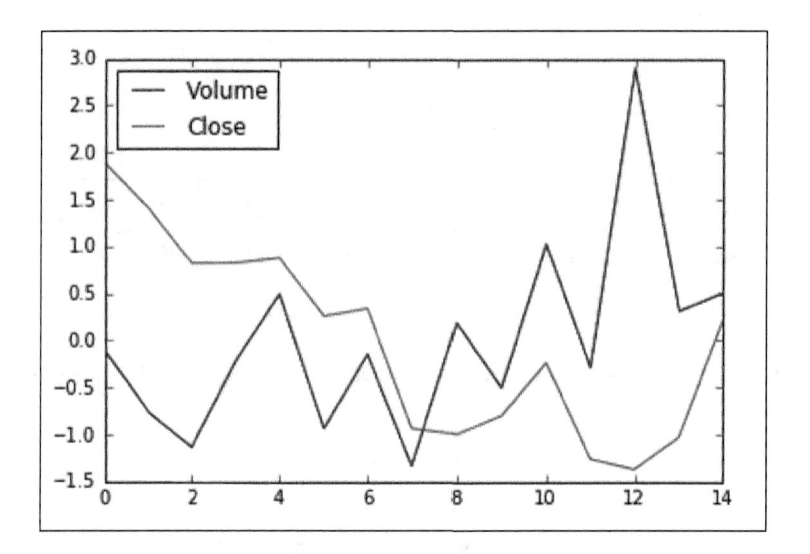

That looks much better! You can see how as the price of AAPL went down somewhere in the middle, the volume of trading also went up! This is actually fairly common:

```
# concatinate prices.Close, and daily_tweets.sentiment

merged = pd.concat([prices.Close, daily_tweets.sentiment], axis=1)
merged.head()
```

	Close	sentiment
Date		
2015-05-02	NaN	0.083031
2015-05-03	NaN	0.107789
2015-05-04	128.699997	0.084062
2015-05-05	125.800003	0.063882
2015-05-06	125.010002	0.066166

Hmm, why are there some null `Close` values? Well if you look up May 2, 2015 on a calendar, you will see that it is a Saturday and the markets are closed on Saturdays, meaning there cannot be a closing price! So we need to make a decision on whether or not to remove these rows because we still have sentiment for that day. Eventually, we will be attempting to predict the next day's closing price and whether the price increased or not so let's go ahead and remove any null values in our dataset:

```
# Delete any rows with missing values in any column
merged.dropna(inplace=True)
```

Now let's attempt to graph our plot:

```
merged.plot()
# wow that looks awful
```

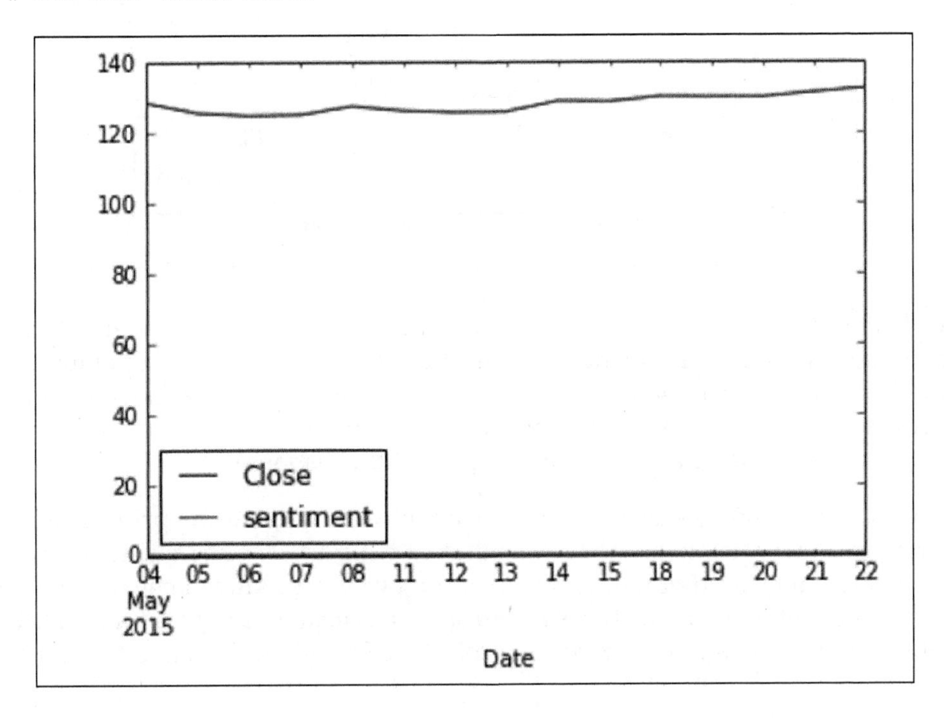

Wow that's terrible. Once again, we must scale our features in order to see any valuable insight:

```
# scale the columns by z scores using StandardScaler
from sklearn.preprocessing import StandardScaler
s = StandardScaler()
```

```
merged_scaled = s.fit_transform(merged)

pd.DataFrame(merged_scaled, columns=merged.columns).plot()
# notice how sentiment seems to follow the closing price
```

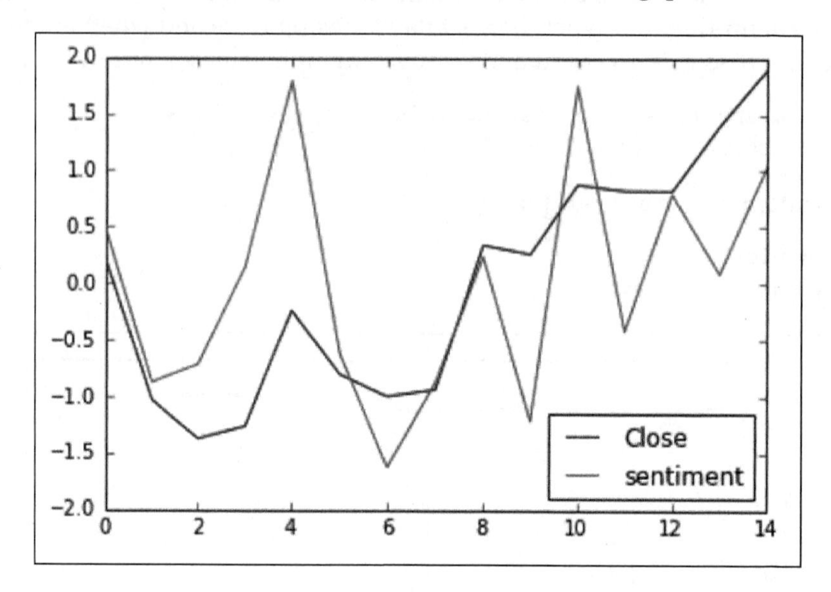

Much better! You can start to see how the closing price of the stock actually does seem to move with our sentiment. Let's take this one step further and attempt to apply a supervised learning model. For this to work, we need to define our features and our response. Recall that our response is the value that we wish to predict and our features are values that we will use to predict the response.

If we look at each row of our data, we have a sentiment and closing price for that day. However, we wish to use today's sentiment to predict *tomorrow's* stock price and whether it increased or not. Think about it; it would be kind of cheating because today's sentiment will include tweets from after the closing price was finalized. To simplify this, we will ignore any tweet as a feature for the prediction of today's price.

So for each row, our response should be today's closing price while our feature should be *yesterday's* sentiment of the stock. To do this, I will use a built-in function in Pandas called `shift` to shift our sentiment column one item backwards:

```
# Shift the sentiment column backwards one item

merged['yesterday_sentiment'] = merged['sentiment'].shift(1)
merged.head()
```

	Close	sentiment	yesterday_sentiment
Date			
2015-05-04	128.699997	0.084062	NaN
2015-05-05	125.800003	0.063882	0.084062
2015-05-06	125.010002	0.066166	0.063882
2015-05-07	125.260002	0.078892	0.066166
2015-05-08	127.620003	0.102898	0.078892

Dataframe with yesterday's sentiment included.

Ah good, now for each day we have our true feature, which is `yesterday_sentiment`. Note that in our head (first five rows) we have a new null value! This is because on the first day we don't have a value from yesterday so we will have to remove it. But before we do, let's define our response column.

We have two options:

- Keep our response quantitative and use a regression analysis
- Convert our response to a qualitative state and use classification

Which route to choose is up to the data scientist and depends on the situation. If you are merely wishing to associate sentiment with a movement in price, then I recommend using the classification route. If you wish to associate sentiment with the amount of movement, I recommend a regression. I will do both!

Regression route

We are already good to go on this front! We have our response and our single feature. We first will have to remove that one null value before continuing:

```
# Make a new dataframe for our regression and drop the null values

regression_df = merged[['yesterday_sentiment', 'Close']]
regression_df.dropna(inplace=True)
regression_df.head()
```

	yesterday_sentiment	Close
Date		
2015-05-05	0.084062	125.800003
2015-05-06	0.063882	125.010002
2015-05-07	0.066166	125.260002
2015-05-08	0.078892	127.620003
2015-05-11	0.102898	126.320000

Let's use both a random forest and a linear regression and see which performs better, using RMSE as our metric:

```
# Imports for our regression

from sklearn.linear_model import LinearRegression
from sklearn.ensemble import RandomForestRegressor
from sklearn.cross_validation import cross_val_score
import numpy as np
```

We will use a cross-validated RMSE in order to compare our two models:

```
# Our RMSE as a result of cross validation linear regression

linreg = LinearRegression()
rmse_cv = np.sqrt(abs(cross_val_score(linreg, regression_
df[['yesterday_sentiment']], regression_df['Close'], cv=3,
scoring='mean_squared_error').mean()))
rmse_cv

3.49837
```

```
# Our RMSE as a result of cross validation random forest

rf = RandomForestRegressor()
rmse_cv = np.sqrt(abs(cross_val_score(rf, regression_df[['yesterday_
sentiment']], regression_df['Close'], cv=3, scoring='mean_squared_
error').mean()))
rmse_cv

3.30603
```

Look at our RMSE, it's about 3.5 for both models, meaning that on average, our model is off by about 3.5 dollars, which is actually a big deal considering our stock price likely doesn't move *that* much:

```
regression_df['Close'].describe()

count      14.000000
mean      128.132858
std         2.471810    # Our standard deviation is less than our RMSE
(bad sign)
min       125.010002
25%       125.905003
50%       128.195003
75%       130.067505
max       132.539993
```

Another way to test the validity of our model is by comparing our RMSE to the null model's RMSE. The null model for a regression model is predicting the average value for each value:

```
# null model for regression
mean_close = regression_df['Close'].mean()
preds = [mean_close]*regression_df.shape[0]
preds
from sklearn.metrics import mean_squared_error
null_rmse = np.sqrt(mean_squared_error(preds, regression_df['Close']))
null_rmse


2.381895
```

Because our model did not beat the null model, maybe regression isn't the best way to go...

Classification route

For classification, we have a bit more work to do because we don't have a categorical response yet. To make one, we need to transform the closing column into some categorical option. I will choose to make the following response. I will make a new column called `change_close_big_deal`, defined as follows:

$$\text{change\_close\_big\_deal} = \begin{cases} 1, & \textit{the change in stock price was} > 1\% \textit{ or } < -1\% \\ 0, & \textit{else} \end{cases}$$

So our response will be *1* if our response changed by a great deal and *0* if the change in stock was negligible:

```
# Imports for our classification

from sklearn.linear_model import LogisticRegression
from sklearn.ensemble import RandomForestClassifier
from sklearn.cross_validation import cross_val_score
import numpy as np



# Make a new dataframe for our classification and drop the null values

classification_df = merged[['yesterday_sentiment', 'Close']]

# variable to represent yesterday's closing price
classification_df['yesterday_close'] = classification_df['Close'].
shift(1)


# column that represents the precent change in price since yesterday
classification_df['percent_change_in_price'] = (classification_
df['Close']-classification_df['yesterday_close']) / classification_
df['yesterday_close']


# drop any null values
classification_df.dropna(inplace=True)
classification_df.head()
```

```
# Our new classification response

classification_df['change_close_big_deal'] = abs(classification_
df['percent_change_in_price'] ) > .01
classification_df.head()
```

	yesterday_sentiment	Close	yesterday_close	percent_change_in_price	change_close_big_deal
Date					
2015-05-05	0.084062	125.800003	128.699997	-0.022533	True
2015-05-06	0.063882	125.010002	125.800003	-0.006280	False
2015-05-07	0.066166	125.260002	125.010002	0.002000	False
2015-05-08	0.078892	127.620003	125.260002	0.018841	True
2015-05-11	0.102898	126.320000	127.620003	-0.010187	True

Our DataFrame with a new column called change_close_big_deal that is either True or False.

Let's now perform the same cross-validation as we did with our regression, instead this time we will be using the *accuracy* feature of our cross-validation module and instead of a regression module we will be using two classification machine learning algorithms:

```
# Our accuracy as a result of cross validation random forest

rf = RandomForestClassifier()
accuracy_cv = cross_val_score(rf, classification_df[['yesterday_
sentiment']], classification_df['change_close_big_deal'], cv=3,
scoring='accuracy').mean()

accuracy_cv

0.1777777
```

Ufff! Not so good, let's try logistic regression though:

```
# Our accuracy as a result of cross validation logistic regression

logreg = LogisticRegression()
accuracy_cv = cross_val_score(logreg, classification_df[['yesterday_
sentiment']], classification_df['change_close_big_deal'], cv=3,
scoring='accuracy').mean()

accuracy_cv

0.5888
```

Better! But of course we should check it with our null model's accuracy:

```
# null model for classification
null_accuracy = 1 - classification_df['change_close_big_deal'].mean()

null_accuracy

0.5833333
```

Whoa, our model can beat the null accuracy meaning that our machine learning algorithm can predict the movement of a stock price using social media sentiment better than just randomly guessing!

Going beyond with this example

There are many ways that we could have enhanced this example to make a more robust prediction. We could have included more features, including a moving average of sentiment instead of looking simply at the previous day's sentiment. We could have also brought in more examples to enhance our idea of sentiment. We could have looked at Facebook, the media, and so on, for more information on how we believe the stock will perform in the future.

We really only had 14 data points, which is *far* from enough to make a production-ready algorithm. Of course for the purposes of this book this is enough, but if we are serious about making a financial algorithm that can effectively predict the stock price movement we will have to obtain many more days of media and prices.

We could have spent more time optimizing for our parameters in our models by utilizing the `gridsearchCV` module in `sklearn` package to get the most out of our models. There are other models that exist that deal specifically with time series data (data that changes over time), including a model called **ARIMA**. Models such as ARIMA and similar ones attempt to focus and zero in on specific time series features.

Case study 2 – why do some people cheat on their spouses?

In 1978, a survey was conducted on housewives in order to discern factors that lead them to pursue extra-marital affairs. This study became the basis for many future studies of both men and women, all attempting to focus on features of people and marriages that led either partner to seek partners elsewhere behind their spouse's back.

Supervised learning is not always about prediction. In this case study, we will purely attempt to identify a few factors of many that we believe might be the most important factors that might lead someone to pursue an affair.

First let's read in the data:

```
# Using dataset of a 1978 survey conducted to measure likliehood of
women to perform extramarital affairs
# http://statsmodels.sourceforge.net/stable/datasets/generated/fair.
html

import statsmodels.api as sm
affairs_df = sm.datasets.fair.load_pandas().data
affairs_df.head()
```

	rate_marriage	age	yrs_married	children	religious	educ	occupation	occupation_husb	affairs
0	3.0	32.0	9.0	3.0	3.0	17.0	2.0	5.0	0.111111
1	3.0	27.0	13.0	3.0	1.0	14.0	3.0	4.0	3.230769
2	4.0	22.0	2.5	0.0	1.0	16.0	3.0	5.0	1.400000
3	4.0	37.0	16.5	4.0	3.0	16.0	5.0	5.0	0.727273
4	5.0	27.0	9.0	1.0	1.0	14.0	3.0	4.0	4.666666

The `statsmodels` website provides a data dictionary, as follows:

- `rate_marriage`: The rating given to the marriage (given by the wife), *1 = very poor, 2 = poor, 3 = fair, 4 = good, 5 = very good*; ordinal level
- `age`: Age of the wife; ratio level
- `yrs_married`: Number of years married: ratio level
- `children`: Number of children between husband and wife: ratio level
- `religious`: How religious the wife is, *1 = not, 2 = mildly, 3 = fairly, 4 = strongly*; ordinal level
- `educ`: Level of education, *9 = grade school, 12 = high school, 14 = some college, 16 = college graduate, 17 = some graduate school, 20 = advanced degree*; ratio level
- `occupation`: *1 = student, 2 = farming, agriculture; semi-skilled, or unskilled worker; 3 = white-collar; 4 = teacher, counselor, social worker, nurse; artist, writers; technician, skilled worker, 5 = managerial, administrative, business, 6 = professional with advanced degree*; nominal level

- `occupation_husb`: Husband's occupation. Same as occupation; nominal level
- `affairs`: Measure of time spent in extramarital affairs; ratio level

Okay, so we have a quantitative response, but my question is simply what factors cause someone to have an affair. The exact number of minutes or hours does not really matter that much. For this reason, let's make a new categorical variable called `affair_binary`, which is either true (they had an affair for more than 0 minutes) or false (they had an affair for 0 minutes):

```
# Create a categorical variable

affairs_df['affair_binary'] = (affairs_df['affairs'] > 0)
```

Again, this column has either a true, or a false value. The value is true if the person had an extra-marital affair for more than 0 minutes. The value is false otherwise. From now on, let's use this binary response as our primary response. Now we are trying to find which of these variables are associated with our response so let's begin.

Let's start with a simple correlation matrix. Recall that this matrix shows us linear correlations between our quantitative variables and our response. I will show the correlation matrix as both a matrix of decimals and also as a heat map. Let's see the numbers first:

```
# find linear correlations between variables and affair_binary
affairs_df.corr()
```

	rate_marriage	age	yrs_married	children	religious	educ	occupation	occupation_husb	affairs	affair_binary
rate_marriage	1.000000	-0.111127	-0.128978	-0.129161	0.078794	0.079869	0.039528	0.027745	-0.178068	-0.331776
age	-0.111127	1.000000	0.894082	0.673902	0.136598	0.027960	0.106127	0.162567	-0.089964	0.146519
yrs_married	-0.128978	0.894082	1.000000	0.772806	0.132683	-0.109058	0.041782	0.128135	-0.087737	0.203109
children	-0.129161	0.673902	0.772806	1.000000	0.141845	-0.141918	-0.015068	0.086660	-0.070278	0.159833
religious	0.078794	0.136598	0.132683	0.141845	1.000000	0.032245	0.035746	0.004061	-0.125933	-0.129299
educ	0.079869	0.027960	-0.109058	-0.141918	0.032245	1.000000	0.382286	0.183932	-0.017740	-0.075280
occupation	0.039528	0.106127	0.041782	-0.015068	0.035746	0.382286	1.000000	0.201156	0.004469	0.028981
occupation_husb	0.027745	0.162567	0.128135	0.086660	0.004061	0.183932	0.201156	1.000000	-0.015614	0.017637
affairs	-0.178068	-0.089964	-0.087737	-0.070278	-0.125933	-0.017740	0.004469	-0.015614	1.000000	0.464046
affair_binary	-0.331776	0.146519	0.203109	0.159833	-0.129299	-0.075280	0.028981	0.017637	0.464046	1.000000

Correlation matrix for marital affairs data from a Likert survey conducted in 1978.

Remember we ignore the diagonal series of 1s because they are merely telling us that every quantitative variable is correlated with itself. Note the other correlated variables, which are the values closest to 1 and -1 on the last row or column (the matrix is always symmetrical across the diagonal).

We can see a few standout variables:

- `affairs`
- `age`
- `yrs_married`
- `children`

These are the top four variables with the largest magnitude (absolute value). However, one of these variables is *cheating*. The `Affairs` variable is the largest in magnitude, but is oviously correlated to `affair_binary` because we made the variable `affair_binary` directly based on affairs. So let's ignore that one. Let's take a look at our correlation heat map to see if our views can be seen there:

```
import seaborn as sns
sns.heatmap(affairs_df.corr())
```

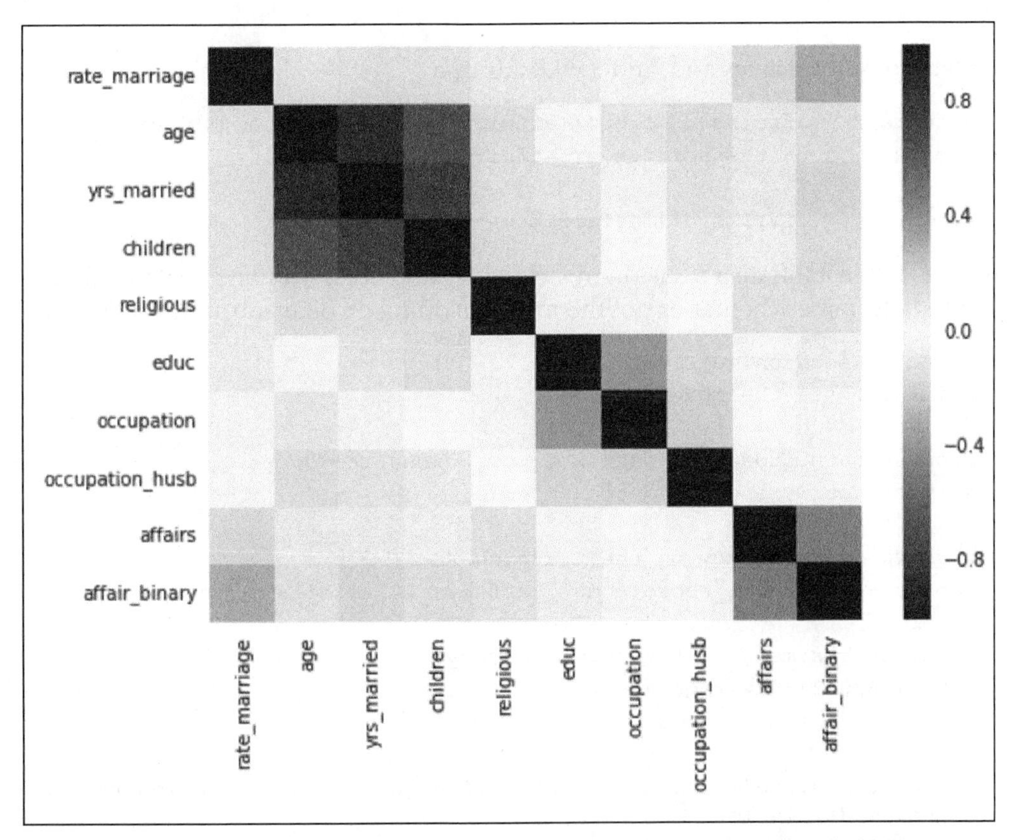

The same correlation matrix, but this time as a heat map. Note the colors close to dark red and dark blue (excluding the diagonal).

We are looking for the dark red and dark blue areas of the heat map. These colors are associated with the most correlated features.

Remember correlations are not the only way to identify which features are associated with our response. This method shows us how linearly correlated the variables are with each other. We may find another variable that affects affairs by evaluating the coefficients of a decision tree classifier. These methods might reveal new variables that are associated with our variables, but not in a linear fashion.

> Also notice that there are two variables here that don't actually belong... Can you spot them? It's the `occupation` and `occupation_husb` variables. Recall earlier we deemed them as nominal and therefore have no right to be included in this correlation matrix. This is because Pandas, unknowingly, casts them as integers and now considers them as quantitative variables. Don't worry, we will fix this soon.

First let's make ourselves an `X` and a `y` Dataframe:

```
affairs_X = affairs_df.drop(['affairs', 'affair_binary'], axis=1)
# data without the affairs or affair_binary column

affairs_y = affairs_df['affair_binary']
```

Now we will instantiate a decision tree classifier and cross-validate our model in order to determine whether or not the model is doing an okay job at fitting our data:

```
model = DecisionTreeClassifier()
# instantiate the model

from sklearn.cross_validation import cross_val_score
# import our cross validation module

# check the accuracy on the training set
scores = cross_val_score(model, affairs_X, affairs_y, cv=10)

print scores.mean(), "average accuracy"
0.659756806845 average accuracy


print scores.std(), "standard deviation" # very low, meaning variance
of the model is low
0.0204081732291 standard deviation

# Looks ok on the cross validation side
```

Because our standard deviation is low we may make the assumption that the variance of our model is low (because variance is the square of standard deviation). This is good because that means that our model is not fitting wildly differently on each fold of the cross-validation and it is generally a reliable model.

Because we agree that our decision tree classifier is generally a reliable model, we can fit the tree to our entire dataset and use the importance metric to identify which variables our tree deemed the most important:

```
# Explore individual features that make the biggest impact
# rate_marriage, yrs_married, and occupation_husb. But one of these
variables doesn't quite make sense right?
# Its the occupation variable, because they are nominal, their
interpretations
model.fit(affairs_X, affairs_y)
pd.DataFrame({'feature':affairs_X.columns, 'importance':model.feature_
importances_}).sort('importance').tail(3)
```

	feature	importance
2	yrs_married	0.136953
0	rate_marriage	0.142588
7	occupation_husb	0.173304

So, `yrs_married` and `rate_marriage` both are important, but the most important variable is `occupation_husb`. But that doesn't make sense because that variable is nominal! So let's apply our dummy variable technique wherein we create new columns that represent each option for `occupation_husb` and also for `occupation`.

For the `occupation` column:

```
# Dummy Variables:

# Encoding qualitiative (nominal) data using separate columns (see
slides for linear regression for more)

occuptation_dummies = pd.get_dummies(affairs_df['occupation'],
prefix='occ_').iloc[:, 1:]

# concatenate the dummy variable columns onto the original DataFrame
(axis=0 means rows, axis=1 means columns)
affairs_df = pd.concat([affairs_df, occuptation_dummies], axis=1)
affairs_df.head()
```

This new Dataframe has many new columns:

educ	occupation	occupation_husb	affairs	affair_binary	occ__2.0	occ__3.0	occ__4.0	occ__5.0	occ__6.0
17.0	2.0	5.0	0.111111	True	1.0	0.0	0.0	0.0	0.0
14.0	3.0	4.0	3.230769	True	0.0	1.0	0.0	0.0	0.0
16.0	3.0	5.0	1.400000	True	0.0	1.0	0.0	0.0	0.0
16.0	5.0	5.0	0.727273	True	0.0	0.0	0.0	1.0	0.0
14.0	3.0	4.0	4.666666	True	0.0	1.0	0.0	0.0	0.0

Remember, these new columns, `occ_2.0`, `occ_4.0`, and so on, represent a binary variable that represents whether or not the wife holds job `2`, or `4`, and so on:

```
# Now for the husband's job

occuptation_dummies = pd.get_dummies(affairs_df['occupation_husb'],
prefix='occ_husb_').iloc[:, 1:]

# concatenate the dummy variable columns onto the original DataFrame
(axis=0 means rows, axis=1 means columns)
affairs_df = pd.concat([affairs_df, occuptation_dummies], axis=1)
affairs_df.head()

(6366, 15)
```

Now we have 15 new columns! Let's run our tree again and find the most important variables:

```
# remove appropiate columns for feature set
affairs_X = affairs_df.drop(['affairs', 'affair_binary', 'occupation',
'occupation_husb'], axis=1)
affairs_y = affairs_df['affair_binary']

model = DecisionTreeClassifier()
from sklearn.cross_validation import cross_val_score
# check the accuracy on the training set
scores = cross_val_score(model, affairs_X, affairs_y, cv=10)
print scores.mean(), "average accuracy"
print scores.std(), "standard deviation" # very low, meaning variance
of the model is low

# Still looks ok
```

```
# Explore individual features that make the biggest impact
model.fit(affairs_X, affairs_y)
pd.DataFrame({'feature':affairs_X.columns, 'importance':model.feature_
importances_}).sort('importance').tail(10)
```

	feature	importance
15	occ_husb__6.0	0.024299
11	occ_husb__2.0	0.030418
14	occ_husb__5.0	0.042021
13	occ_husb__4.0	0.047874
4	religious	0.098630
1	age	0.111628
5	educ	0.131468
2	yrs_married	0.132034
3	children	0.134374
0	rate_marriage	0.139502

- `age`
- `yrs_married`
- `children`

And there you have it:

- `rate_marriage`: the rating of the marriage, as told by the decision tree
- `children`: the number of children they had, as told by the decision tree and our correlation matrix
- `yrs_married`: the number of years they had been married, as told by the decision tree and our correlation matrix
- `educ`: the level of education the women had, as told by the decision tree
- `age`: the age of the women, as told by the decision tree and our correlation matrix

These seem to be the top five most important variables in determining whether or not a woman from the 1978 survey would be involved in an extramarital affair.

Case study 3 – using tensorflow

I would like to finish off our time together by looking at a somewhat more modern module that was only recently introduced by Google's machine learning division called **tensorflow**.

Tensorflow is an open-source machine learning module that is used primarily for its simplified deep learning and neural network abilities. I would like to take some time to introduce the module and solve a few quick problems using tensorflow. The syntax for tensorflow (like PyBrain in *Chapter 12, Beyond the Essentials*) is a bit different than our normal *scikit-learn* syntax so I will be going over it step by step. Let's start with some imports:

```
from sklearn import datasets, metrics
import tensorflow as tf
import numpy as np
from sklearn.cross_validation import train_test_split
%matplotlib inline
```

Our imports from `sklearn` include `train_test_split`, `datasets`, and `metrics`. We will be utilizing our train-test splits to reduce overfitting, we will use datasets in order to import our `iris` classification data and we'll use the metrics module in order to calculate some simple metrics for our learning models.

Tensorflow learns in a different way in that it is always trying to minimize an error function. It does this by iteratively going through our entire dataset and every so often, updates our model to better fit the data.

It is important to note that tensorflow doesn't just implement neural networks, but it can implement even simpler models as well. For example, let's implement a classic logistic regression using tensorflow:

```
# Our data set of iris flowers
iris = datasets.load_iris()

# Load datasets and split them for training and testing
X_train, X_test, y_train, y_test = train_test_split(iris.data, iris.
target)


####### TENSORFLOW #######

# Here is tensorflow's syntax for defining features.
```

```
# We must specify that all features have real-value data
feature_columns = [tf.contrib.layers.real_valued_column("",
dimension=4)]
# notice the dimension is set to four because we have four columns

# We set our "learning rate" which is a decimal that tells the network
# how quickly to learn
optimizer = tf.train.GradientDescentOptimizer(learning_rate=.1)
# A learning rate closer to 0 means the network will learn slower


# Build a linear classifier (logistic regression)
# note we have to tell tensorflow the number of classes we are looking
for
# which are 3 classes of iris
classifier = tf.contrib.learn.LinearClassifier(feature_
columns=feature_columns,
                                               optimizer=optimizer,
                                                      n_classes=3)

# Fit model. Uses error optimization techniques like stochastic
gradient descent
classifier.fit(x=X_train,
               y=y_train,
               steps=1000)  # number of iterations
```

I will point out the key lines of code from the preceding snippet to really solidify what is happening during training:

- **feature_columns = [tf.contrib.layers.real_valued_column("", dimension=4)]**

  Here I am creating four input *columns* that we know correlate to the flowers' sepal length, sepal width, petal length, and petal width.

- **optimizer = tf.train.GradientDescentOptimizer(learning_rate=.1)**

  Here I am telling tensorflow to optimize using something called **gradient descent**, which means that we will define an error function (which will happen in the next step) and little by little, we will work our way to minimize this error function.

  Our learning rate should hover close to 0 because we want our model to learn slowly. If our model learns too quickly it might "skip over" the right answer!

- **classifier = tf.contrib.learn.LinearClassifier(feature_columns=feature_columns, optimizer=optimizer, n_classes=3)**:

  When we specify `LinearClassifier` we are denoting the same error function that logistic regression is minimizing, meaning that this classifier is attempting to work as a logistic regression classifier.

  We give the model our `feature_columns` as defined in step 1.

  The `optimizer` is the method of minimizing our error function; in this case we chose gradient descent.

  We also must specify our number of classes as being `3`. We know that we have three different iris flowers that the model could choose from.

- **classifier.fit(x=X_train, y=y_train, steps=1000)**:

  The train looks similar to a scikit-learn model with an added parameter called `steps`. Steps tell us how many times we would like to go over our dataset. So when we specify `1000` we are iterating over our dataset. The more steps we take, the more the model gets a chance to learn.

Phew! When we run the preceding code, a linear classifier (logistic regression) model is being fit and when it is done it is ready to be tested:

```
# Evaluate accuracy.
accuracy_score = classifier.evaluate(x=X_test,
                                     y=y_test)["accuracy"]

print('Accuracy: {0:f}'.format(accuracy_score))
Accuracy: 0.973684
```

Excellent! It is worth noting that when using tensorflow, we may also utilize a similarly simple, `predict` function:

```
# Classify two new flower samples.
new_samples = np.array(
    [[6.4, 3.2, 4.5, 1.5], [5.8, 3.1, 5.0, 1.7]], dtype=float)

y = classifier.predict(new_samples)
print('Predictions: {}'.format(str(y)))
Predictions: [1 2]
```

Now let's compare this with a standard scikit-learn logistic regression to see who won:

```
from sklearn.linear_model import LogisticRegression
# compare our result above to a simple scikit-learn logistic
regression

logreg = LogisticRegression()
# instantiate the model

logreg.fit(X_train, y_train)
# fit it to our training set

y_predicted = logreg.predict(X_test)
# predict on our test set, to avoid overfitting!


accuracy = metrics.accuracy_score(y_predicted, y_test)
# get our accuracy score

accuracy
# It's the same thing!
```

Wow, so it seems that with a 1,000 steps, a gradient descent optimized tensorflow model is no better than a simple sklearn logistic regression. OK, that's fine, but what if we allowed the model to iterate over the `iris` dataset even more?

```
feature_columns = [tf.contrib.layers.real_valued_column("",
dimension=4)]

optimizer = tf.train.GradientDescentOptimizer(learning_rate=.1)

classifier = tf.contrib.learn.LinearClassifier(feature_
columns=feature_columns,
                                               optimizer=optimizer,
                                            n_classes=3)


classifier.fit(x=X_train,
               y=y_train,
               steps=2000)  # number of iterations is 2000 now
```

Our code is exactly the same as before, but now we have `2000` steps instead of `1000`:

```
# Evaluate accuracy.
accuracy_score = classifier.evaluate(x=X_test,
                                     y=y_test)["accuracy"]


print('Accuracy: {0:f}'.format(accuracy_score))
Accuracy: 0.973684
```

And now we have an even better accuracy!

Note that you need to be very careful in choosing the number of steps. As you increase this number, you increase the number of times your model sees the same exact training points over and over again. We do have a chance of becoming overfit! To remedy this, I would recommend choosing multiple train test splits and running the model on each one (**k-fold cross-validation**).

It is also worth mentioning that tensorflow implements very low bias high variance models. Meaning that running the preceding code again for tensorflow might result in a different answer! This is one of the caveats of deep learning. They might converge to a very great low bias model, but that model will have a high variance and therefore, amazingly, might not generalize to all of the sample data. As mentioned before, a cross-validation would be helpful in order to mitigate this.

Tensorflow and neural networks

Now let's point a more powerful model at our `iris` dataset. Let's create a neural network who's goal it is to classify iris flowers (because why not?):

```
# Specify that all features have real-value data
feature_columns = [tf.contrib.layers.real_valued_column("",
dimension=4)]

optimizer = tf.train.GradientDescentOptimizer(learning_rate=.1)


# Build 3 layer DNN with 10, 20, 10 units respectively.
classifier = tf.contrib.learn.DNNClassifier(feature_columns=feature_
columns,
```

```
                                         hidden_units=[10, 20, 10],
                                         optimizer=optimizer,
                                         n_classes=3)

# Fit model.
classifier.fit(x=X_train,
               y=y_train,
               steps=2000)
```

Notice that our code really hasn't changed from the last segment. We still have our `feature_columns` from before, but now we introduce, instead of a linear classifier, a **DNNClassifier**, which stands for **Deep Neural Network Classifier**.

This is tensorflow's syntax for implementing a neural network. Let's take a closer look:

```
tf.contrib.learn.DNNClassifier(feature_columns=feature_columns,
                                         hidden_units=[10, 20, 10],
                                         optimizer=optimizer,
                                         n_classes=3)
```

We see that we are inputting the same `feature_columns`, `n_classes`, and `optimizer`, but see how we have a new parameter called `hidden_units`? This list represents the number of nodes to have in each layer between the input and the output layer.

All in all, this neural network will have five layers:

- The first layer will have four nodes, one for each of the iris feature variables. This layer is the input layer.
- A hidden layer of 10 nodes.
- A hidden layer of 20 nodes.
- A hidden layer of 10 nodes.
- The final layer will have three nodes, one for each possible outcome of the network. This is called our output layer.

Now that we've trained our model, let's evaluate it on our test set:

```
# Evaluate accuracy.
accuracy_score = classifier.evaluate(x=X_test,
                                     y=y_test)["accuracy"]
print('Accuracy: {0:f}'.format(accuracy_score))
Accuracy: 0.921053
```

Hmm, our neural network didn't do so well on this dataset, but perhaps it is because the network is a bit too complicated for such a simple dataset. Let's introduce a new dataset that has a bit more to it...

The `MNIST` dataset consists of over 50,000 handwritten digits (0-9) and the goal is to recognize the handwritten digits and output which letter they are writing. Tensorflow has a built-in mechanism for downloading and loading these images. We've seen these images before, but at a much smaller scale in *Chapter 12, Beyond the Essentials*:

```
from tensorflow.examples.tutorials.mnist import input_data
mnist = input_data.read_data_sets("MNIST_data/", one_hot=False)

Extracting MNIST_data/train-images-idx3-ubyte.gz
Extracting MNIST_data/train-labels-idx1-ubyte.gz
Extracting MNIST_data/t10k-images-idx3-ubyte.gz
Extracting MNIST_data/t10k-labels-idx1-ubyte.gz
```

Notice that one of our inputs for downloading `mnist` is called `one_hot`. This parameter either brings in the dataset's target variable (which is the digit itself) as a single number or has a dummy variable.

For example, if the first digit were a 7, the target would either be:

- 7: If `one_hot` was false
- 0 0 0 0 0 0 0 1 0 0: If `one_hot` was true (notice that starting from 0, the seventh index is a 1)

We will encode our target the former way, as this is what our tensorflow neural network and our sklearn logistic regression will expect.

The dataset is split up already into a training and test set, so let's create new variables to hold them:

```
x_mnist = mnist.train.images
y_mnist = mnist.train.labels.astype(int)
```

For the `y_mnist` variable, I specifically cast every target as an integer (by default they come in as floats) because otherwise tensorflow would throw an error at us.

Out of curiosity, let's take a look at a single image:

```
import matplotlib.pyplot as plt
plt.imshow(x_mnist[10].reshape(28, 28))
```

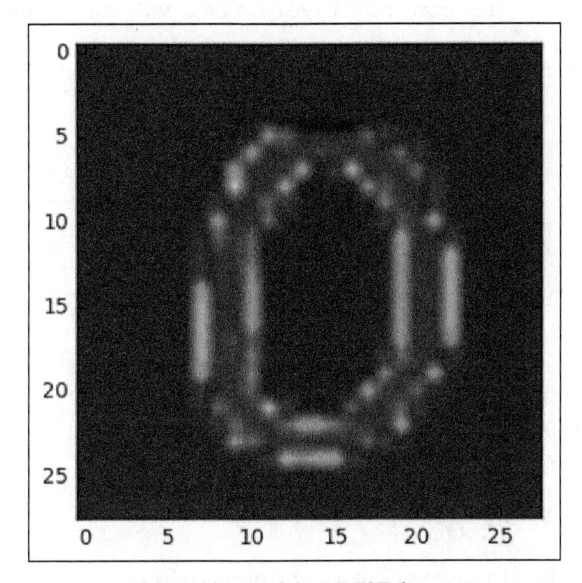

The number 0 of the MNIST dataset

And hopefully our target variable matches at the 10th index as well:

```
y_mnist[10]
0
```

Excellent! Let's now take a peek at how big our dataset is:

```
x_mnist.shape
(55000, 784)

y_mnist.shape
(55000,)
```

Our training size then is `55000` images and target variables.

Let's fit a deep neural network to our images and see if it will be able to pick up on the patterns in our inputs:

```
# Specify that all features have real-value data
feature_columns = [tf.contrib.layers.real_valued_column("",
dimension=784)]
optimizer = tf.train.GradientDescentOptimizer(learning_rate=.1)


# Build 3 layer DNN with 10, 20, 10 units respectively.
classifier = tf.contrib.learn.DNNClassifier(feature_columns=feature_
columns,
                                            hidden_units=[10, 20, 10],
                                               optimizer=optimizer,
                                                   n_classes=10)


# Fit model.
classifier.fit(x=x_mnist,
               y=y_mnist,
               steps=1000)
# Warning this is veryyyyyyyy slow
```

This code is very similar to our previous segment using `DNNClassifier`; however, look how in our first line of code, I have changed the number of columns to be `784` while in the classifier itself, I changed the number of output classes to be `10`. These are manual inputs that tensorflow must be given to work.

The preceding code runs *very* slowly. It is little by little adjusting itself in order to get the best possible performance from our training set. Of course, we know that the ultimate test here is testing our network on an unknown test set, which is also given to us from tensorflow:

```
x_mnist_test = mnist.test.images
y_mnist_test = mnist.test.labels.astype(int)

x_mnist_test.shape
(10000, 784)


y_mnist_test.shape
(10000,)
```

So we have 10,000 images to test on; let's see how our network was able to adapt to the dataset:

```
# Evaluate accuracy.
accuracy_score = classifier.evaluate(x=x_mnist_test,
                                     y=y_mnist_test)["accuracy"]
print('Accuracy: {0:f}'.format(accuracy_score))
Accuracy: 0.920600

Not bad, 92% accuracy on our dataset. Let's take a second and compare
this performance to a standard sklearn logistic regression now:
logreg = LogisticRegression()
logreg.fit(x_mnist, y_mnist)
# Warning this is slow

y_predicted = logreg.predict(x_mnist_test)
from sklearn.metrics import accuracy_score
# predict on our test set, to avoid overfitting!

accuracy = accuracy_score(y_predicted, y_mnist_test)
# get our accuracy score

accuracy
0.91969
```

Success! Our neural network performed better than the standard logistic regression. This is likely because the network is attempting to find relationships between the pixels themselves and using these relationships to map them to what digit we are writing down. In logistic regression, the model assumes that every single input is independent of one another, and therefore has a tough time finding relationships between them.

There are ways of making our neural network learn differently:

- We could make our network wider, that is, increase the number of nodes in the hidden layers instead of having several layers of a smaller number of nodes:

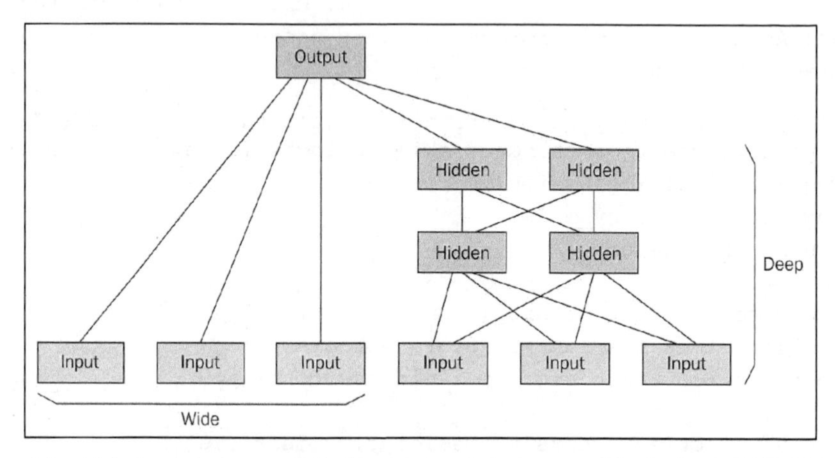

Source: http://electronicdesign.com/site-files/electronicdesign.com/files/uploads/2015/02/0816_Development_Tools_F1_0.gif

```
# A wider network
feature_columns = [tf.contrib.layers.real_valued_column("",
dimension=784)]

optimizer = tf.train.GradientDescentOptimizer(learning_rate=.1)


# Build 3 layer DNN with 10, 20, 10 units respectively.
classifier = tf.contrib.learn.DNNClassifier(feature_
columns=feature_columns,
                                            hidden_units=[1500],
                                            optimizer=optimizer,
                                                   n_classes=10)


# Fit model.
classifier.fit(x=x_mnist,
               y=y_mnist,
               steps=100)
# Warning this is veryyyyyyyy slow
# Evaluate accuracy.
```

```
accuracy_score = classifier.evaluate(x=x_mnist_test,
                                     y=y_mnist_test)["accuracy"]
print('Accuracy: {0:f}'.format(accuracy_score))
      Accuracy: 0.898400
```

- We could increase our learning rate, forcing the network to attempt to converge into an answer faster. As I mentioned before, we run the risk of the model skipping the answer entirely if we go down this route. It is usually better to stick with a smaller learning rate.
- We can change the method of optimization. Gradient descent is very popular; however, there are other algorithms for doing so. One example is called the **Adam Optimizer**. The difference is in the way they traverse the error function, and therefore the way that they approach the optimization point. Different problems in different domains call for different optimizers.
- There is no replacement for a good old fashioned feature selection phase instead of attempting to let the network figure everything out for us. We can take the time to find relevant and meaningful features that actually will allow our network to find an answer quicker!

Summary

In this chapter, we've seen three different case studies from three different domains using many different statistical and machine learning methods. However, what all of them have in common is that in order to solve them properly, we had to implement a data science mindset. We had to solve problems in an interesting way, obtain data, clean the data, visualize the data, and finally, model the data and evaluate our thinking process.

I do hope that you have found the contents of this book to be interesting and not just the final chapter! I leave it unto you to keep exploring the world of data science. Keep learning Python. Keep learning statistics and probability. Keep your minds open. It is my hope that this book has been a catalyst for you to go out and find even more on the subject.

For further readings past this book, I highly recommend looking into well-known data science books and blogs, such as:

- *Dataschool.io*—blog by Kevin Markham
- *Python for Data Scientists* by Packt

If you would like to contact me for any reason, please feel free to reach out to `sinan.u.ozdemir@gmail.com`.

Index

A

A/B test 134
Adam Optimizer 361
addition rule 97
alternative hypothesis 168
anomaly detection 319
ARIMA 342
arithmetic mean 40, 138
arithmetic symbols
  about 69
  dot product 70-73
  proportional 70
  summation 69, 70

B

back-propagation 320
bar charts 185-187
basic Python, example
  about 12
  single Tweet, parsing 13, 14
Bayes formula 70
Bayes theorem
  about 105-107
  applications 110
  examples 108-110
  medical studies example 112, 113
  titanic dataset 110, 111
bias variance tradeoff
  about 290
  error, due to bias 290
  error, due to variance 290-298
  extreme cases 298
  working, with error functions 299-301
big data 16
bi-modal 162
binary classifier 102
binomial random variable
  about 119, 120
  blood types example 121, 122
  restaurant openings example 120, 121
box plots
  about 189
  creating 189-191

C

Cartesian graph 73
causation
  versus correlation 192-194
central limit theorem 164
centroid 263
chi-square goodness of fit test
  about 175
  assumptions 175, 176
  example 176, 177
chi-square test for association/independence
  about 177, 178
  assumptions 178, 179
classification 210
classification tree
  fitting 256-261
cluster 263
clustering 212
coefficient of variation
  about 144
  employee salaries example 144
collectively exhaustive 105

collectively exhaustive events
about 105
examples 106
communication 181, 182
complementary events 100-102
compound events
about 93
example 94-96
conditional probability 96
confidence 164
confidence intervals 164-167
confounding factor 137
confusion matrix 102
continuous data 34
continuous random variable 125-128
correlation
versus causation 192-194
correlation coefficients 151, 152
cross validation error
versus training error visualization 308-310
CSV (comma separated value) 32

D

data
levels 35
organized data 4
types 25, 26
unorganized data 4
data exploration
about 49
basic questions 50
titanic dataset 60-64
yelp dataset 51-53
data mining 15
data model 8
data, obtaining
about 133
experimental 133-135
observational 133
data points 264-269
data preprocessing
example 27
relative length of text 29
topics, picking 29
word/phrase counts 28
data sampling
about 136
probability sampling 136
random sampling 136, 137
unequal probability sampling 137
data science
about 3, 4, 7, 47
data, exploring 48
data, modeling 49
data, obtaining 48
interesting question, asking 48
need for 5
results, communicating 49
results, visualizing 49
Sigma Technologies example 5
steps 48
data science, case studies
about 16
government paper pushing automation 16-18
job descriptions 20-22
marketing dollars 18, 19
data science Venn diagram
about 6, 7
computer programming 6, 10
domain knowledge 6, 14
math/statistics 6, 8
decision trees
about 254-256
versus random forests 317
Deep Neural Network Classifier (DNNClassifier) 355
dimension reduction
about 212
cons 287
discrete data 34
discrete random variables
about 114-119
binomial random variable 119, 120
continuous random variable 125-128
geometric random variable 123
Poisson random variable 124
types 119

domain knowledge 14
dot product 70-73
dummy variables 239-243

E

Empirical rule
about 153, 154
example 154
ENIAC 2
ensembling techniques 310-312
entity movement 319
entropy 256
error functions 299
Euler's number 232
event 88
exploration tips, for qualitative data
about 54
filtering 56, 58
nominal level columns 54, 56
ordinal level columns 58, 59
exploratory data analysis (EDA) 15
exponent
about 74
examples 75, 76
extra-marital affairs case study 342-349
extreme cases, bias/variance tradeoff
overfitting 299
underfitting 298

F

false negative 102, 174
false positive 102, 174
feature extraction
about 275-286
pros 287
feature selection 279
filtering 56
Frequentist approach
about 90
law of large numbers 91, 93
marketing stats example 91

G

geometric random variable
about 122
weather example 123
gini index 256
global score 274
graphs
about 73, 74
bar charts 185, 187
box plots 189-191
Cartesian graph 73
histograms 187, 188
line graphs 184, 185
scatter plots 182-184
grid searching 305-307

H

histograms
about 187
plotting 188
hypothesis test
about 168
conducting 169
one sample t-tests 170
type I error 174
type II error 174
hypothesis test, for categorical values
about 174
chi-square goodness of fit test 175
chi-square test for association/independence 177, 178

I

independent events
examples 100
intersection 79
interval level, of data
about 39
example 39
mathematical operations 40
measures of center 40, 41
measures of variation 41

J

jaccard measure 80

K

K folds cross-validation
about 301, 302
features 302-304
K-means clustering
about 262-264
example 270-272
K-Nearest Neighbors (KNN) algorithm 300
KPI (key performance indicator) 197

L

labeled data 206
levels, data
interval 39
nominal 35
ordinal 36
ratio 43
likelihood 246
likert scale 117
linear algebra
about 73, 81
matrix multiplication 81
linear regression
about 217-222
predictors, adding 222-224
line graphs 184, 185
logarithm
about 75
examples 75
logistic regression 231-238
log odds 236

M

machine learning
about 7, 9, 15, 202
facial recognition example 203
limitations 204
overview 215, 216
supervised learning 206
types 205
unsupervised learning 212, 213
working 205
magnitude 78
margin of error 165
mathematics 65
Math & Statistics Knowledge base 7
matrices
multiplying 82-84
matrix 67, 68
matrix multiplication 81
measures of center 138, 139
measures of relative standing
about 145-150
correlations, in data 151, 152
measures of variation 139-145
median 38, 139
model coefficients 220
models 8
multilayer perceptrons (MLP) 320
multiplication rule 99, 100
mutual exclusivity 98
mutually exhaustive 106

N

Naïve Bayes classification 245-253
neural networks
about 318
advantage 318
basic structure 318-324
nominal level, of data
about 35, 36
mathematical operations 36
measure of center 36
normalizing constant 246
notation 88
null hypothesis 168
null model 229
null set 78

O

odds 233, 234
one sample t-test
about 170
assumptions 171-174
example 170, 171
one-tailed test 172
optimal number
selecting, for cluster validation 273
ordinal level, of data
about 36, 39
examples 37
mathematical operations 37
measures of center 38, 39
organized data 4
overfitting 227, 299

P

parameter 132
pattern recognition 318
perceptron 319
point estimates 157-162
Poisson distribution 158
Poisson random variable
about 124, 158
call center example 125
examples 124
population 131
posterior 245
prediction 208
predictive analytics models 206
presentation, to formal audience
tips 197, 198
Principal Component Analysis (PCA) 276-286
prior probability 245
probabilistic model 15
probability 88, 89, 233
probability density function (PDF) 126
probability mass function (PMF) 120
probability, rules
about 97
addition rule 97
complementary events 100-102
independence 100
multiplication rule 99, 100
mutual exclusivity 98
probability sampling 136
procedure 88
proportional 70
p-value 168
Python
need for 10
practices 11, 12

Q

qualitative data 30
qualitative data, versus quantitative data
about 30
coffee shop data example 30-32
world alcohol consumption data example 32-34
quantitative data
about 30
continuous data 34
discrete data 34

R

random forests
about 312-317
advantages 317
disadvantages 317
versus decision trees 317
random sampling 136, 137
random variables
about 105, 113, 114
discrete random variable 114-119
ratio level, of data
about 43
examples 43
issues 44
measures of center 43, 44

regression 210
regression metrics 224-230
regression tree
  building 256
reinforcement learning
  about 214-216
  cons 216
  pros 216
relative frequency 90
relative length 29

S

sample 132
sample space 88
sampling bias 136
sampling distributions 162-164
scalar 70
scatter plot 182-184
set 77
set theory 77-79
Silhouette Coefficient 273-275
Simpson's paradox 195, 196
slope 74
spawner-recruit models
  example 8, 9
square matrix 67
standard deviation 41, 42, 140
standard normal distribution 126
statistical model 15
statistical modeling 217
statistics 131, 132
statistics, measuring
  about 138
  measure of relative standing 145-150
  measures of center 138, 139
  measures of variation 139-143
stock prices prediction based on social media case study
  about 325
  classification route 340-342
  example 342
  exploratory data analysis 326-337
  regression route 337-339
  text sentiment analysis 325, 326
structured data
  about 26
  versus unstructured data 27
subset 78
Substantive Expertise 7
summation 69, 70
sum of squared residuals 220
supervised learning
  about 206
  classification 210
  cons 215
  example 207, 208
  predictions 209
  pros 215
  regression 210
  types 209
  working 206

T

tensorflow case study
  about 350-354
  neural networks, creating 354-361
test statistic 173
titanic dataset 60-64
training error visualization
  versus cross validation error 308-310
true negatives 102
true positives 102
type I error 174
type II error 174

U

underfitting 298
unequal probability sampling 137
union 79
unorganized data 4
unstructured data
  about 26
  versus structured data 27

unsupervised learning
about 212-216, 262
cons 216
pros 216
reinforcement learning 214
using 262

vector 66, 67
verbal communication
about 196
story telling 197

W

why/how/what strategy, of presentation 198, 199

yelp dataset
Dataframe 53
Series object 54

Z

z-score 145